TRANSPORTATION INFRASTRUCTURE - ROADS, HIGHWAYS, BRIDGES, AIRPORTS AND MASS TRANSIT

MANAGED MOTORWAYS

TRAFFIC OPTIMIZATION AND ROAD PRICING APPROACHES IN EUROPE AND SINGAPORE

Transportation Infrastructure - Roads, Highways, Bridges, Airports and Mass Transit

Additional books in this series can be found on Nova's website under the Series tab.

Additional E-books in this series can be found on Nova's website under the E-book tab.

Transportation Issues, Policies and R&D

Additional books in this series can be found on Nova's website under the Series tab.

Additional E-books in this series can be found on Nova's website under the E-book tab.

TRANSPORTATION INFRASTRUCTURE - ROADS, HIGHWAYS, BRIDGES,
AIRPORTS AND MASS TRANSIT

MANAGED MOTORWAYS

TRAFFIC OPTIMIZATION AND ROAD PRICING APPROACHES IN EUROPE AND SINGAPORE

ELVIS L. DUNCAN
AND
MICHAEL J. SNYDER
EDITORS

nova publishers

Nova Science Publishers, Inc.
New York

For permission to use material from this book please contact us:
Telephone 631-231-7269; Fax 631-231-8175
Web Site: http://www.novapublishers.com

Additional color graphics may be available in the e-book version of this book.

Library of Congress Cataloging-in-Publication Data

ISBN: 978-1-61942-881-2

Published by Nova Science Publishers, Inc. + New York

CONTENTS

PREFACE

Finding cost-effective options to mitigate recurrent and non-recurrent congestion on freeway facilities is one of the most significant challenges State and regional transportation organizations face. Several countries are implementing managed motorway concepts to move higher traffic volumes on their highways more efficiently without acquiring more land and constructing large-scale infrastructure projects. This book explores managed motorway concepts and traffic optimization technologies being implemented in Europe and Singapore.

Chapter 1- Continued growth in travel along congested freeway corridors exceeds the ability of agencies to provide sufficient solutions and alternatives based on traditional roadway expansion and improvement projects. High construction costs, constrained right-of-way, statutory restrictions, and environmental factors are pushing agencies to explore solutions such as active traffic management and managed lanes, which improve safety by reducing collisions and nonrecurring congestion and maximize throughput under congested conditions. Finding cost-effective options to mitigate recurrent and nonrecurrent congestion on freeway facilities is one of the most significant challenges State and regional transportation organizations face.

Chapter 2- The evidence in the countries visited in this scan shows that road pricing can play a vital role in creating new funding for transportation, encouraging improved quality of life in the urban environment, advancing economic productivity for goods movement and business, increasing the use of public transit, and reducing congestion and emissions.

In: Managed Motorways
Editors: Elvis L. Duncan and Michael J. Snyder
ISBN: 978-1-61942-881-2

Chapter 1

FREEWAY GEOMETRIC DESIGN FOR ACTIVE TRAFFIC MANAGEMENT IN EUROPE*

United States Department of Transportation

EXECUTIVE SUMMARY

Introduction

Continued growth in travel along congested freeway corridors exceeds the ability of agencies to provide sufficient solutions and alternatives based on traditional roadway expansion and improvement projects. High construction costs, constrained right-of-way, statutory restrictions, and environmental factors are pushing agencies to explore solutions such as active traffic management and managed lanes, which improve safety by reducing collisions and nonrecurring congestion and maximize throughput under congested conditions. Finding cost-effective options to mitigate recurrent and nonrecurrent congestion on freeway facilities is one of the most significant challenges State and regional transportation organizations face.

Several countries are implementing managed motorway concepts to move higher traffic volumes on their highways more efficiently without acquiring more land and constructing large-scale infrastructure projects. Managed motorway concepts introduce new and revised operational activities that place greater reliance on technology than traditional roadway projects. Managed motorways combine actively or dynamically managed operational regimes, specific infrastructure designs, and technology solutions. They use a range of traffic management measures to actively monitor the motorway based on real-time conditions:

- Dynamically control speeds (see figure 1).
- Add capacity (figure 2, see next page).
- Inform road users of conditions on the network (figure 3, see next page).

* This is an edited, reformatted and augmented version of the U.S. Department of Transportation, Federal Highway Administration report, dated March 2011.

The objective of implementing this range of measures is to optimize traffic and safety performance.

Examples of these measures include shoulder running, variable speed limits, lane control signals, dynamic rerouting, and the provision of driver information using variable message signs. Managed motorway concepts applied in Europe have been proven to reduce collisions, improve journey time reliability, and increase vehicular throughput.

Figure 1. England: variable speed limit in Birmingham (vehicles enter the roadway from the left, opposing traffic is on the right, and speeds shown are in miles per hour).

Figure 2. Netherlands: shoulder running.

Figure 3. England: variable message signs in Birmingham (vehicles enter the roadway from the left, opposing traffic is on the right, and speeds shown are in miles per hour).

Background

In 2006 a scan team observed that transportation agencies in Denmark, England, Germany, and the Netherlands, through the deployment of congestion management strategies, were able to optimize the investment in infrastructure to meet drivers' needs. Strategies included speed harmonization, temporary shoulder use, and dynamic signing and rerouting. The team's recommendations for U.S. implementation included promoting active traffic management to optimize existing infrastructure during recurrent and nonrecurrent congestion, emphasizing customer orientation, focusing on trip reliability, providing consistent messages to roadway users, and making operations a priority in planning, programming, and funding processes.

Since the 2006 scanning study, active traffic management concepts have been implemented in Washington and Minnesota and are being considered in Virginia. During these implementations, several geometric design-related questions were voiced. A scanning study was proposed to obtain a better appreciation for how geometric design is being handled with active traffic management programs. The desk scan revealed that several European countries have implemented innovative geometric design solutions in their active traffic management programs. In June 2010 a team of 10 U.S. transportation professionals with expertise in planning, design, and operation of freeways visited four countries in Europe: England, Germany, the Netherlands, and Spain. The purpose of the scanning study was to examine active traffic management design practices used in other countries to improve the operational performance of congested freeway facilities without compromising safety. This 2010 scan built on other scans that focused on congestion management and managed lane programs.

Key Findings

Key findings from the 2010 scan include the following:

- Much like the United States, many European nations face growing traffic and congestion levels on their freeway networks. Several European highway agencies are responding to growing congestion by implementing active traffic management systems that better use the existing roadway footprint. In Europe, "managed motorways" is the term used to describe the range of traffic management measures implemented to improve traffic flow, enhance safety, and inform road users of conditions on the freeway network. Managed motorway concepts have had great success in the countries the scan team visited, and these strategies and techniques are likely to provide great benefit if applied in the United States.
- The European countries visited comprehensively integrate a suite of complementary techniques to dynamically manage traffic flow in response to changing volumes, speeds, and incidents. The result is demonstrably improved safety, travel time reliability, and congestion relief on urban motorway sections. Techniques that integrate roadway design with operational strategies include the following:

- variable speed limits, line control, and speed harmonization (see figures 1, 4, and 5)
- Shoulder running (figures 2 and 4) with emergency refuge areas (figure 6)
- Queue warning and variable messaging (figure 3)
- 24/7 monitoring of traffic with cameras and/or in-pavement sensors (both to detect incidents and identify when to reduce speed limits) (figure 7, see next page)
- Incident management (figure 8, see next page)
- Automated enforcement (see figure 9 on page 5 for examples of signs)
- Specialized algorithms for temporary shoulder running, variable speed limits, and/ or incident detection and management
- Ramp metering (coordinated or independent function)

- Managed motorway strategies are synergistic and are most effective when applied in an integrated and dynamic system.
- Many managed motorway concepts are applicable to all U.S. metro areas and rural high-volume freeway corridors. The management strategies appropriate for a freeway corridor evolve as the needs and demands of the area change. In other words, transportation officials should recognize that freeways need a continuum of operational and management strategies that change as traffic needs and demands change.
- European countries faced safety concerns similar to those in the United States and successfully addressed those concerns in managed motorway deployments. Managed motorways have contributed to substantial safety improvements in Europe.

Figure 4. Germany: shoulder use and variable speed limit in Hessen (speeds are in kilometers per hour)[11] .

Figure 5. Netherlands: variable speed limit speeds are in kilometers per hour).

Figure 6. England: emergency refuge area in Birmingham (traffic travels on the left side on England's roadways).

Figure 7. Netherlands: surveillance camera and loop detectors.

Figure 8. England: incident management in Birmingham.

- Many European countries went through a paradigm shift in their design policies and practices by adopting risk- and performance- based approaches to making design choices on actively managed freeway facilities. An example of changed design philosophy is considering the dynamic operating regimes of a managed freeway rather than selecting design criteria based on a static operating condition. Successful active traffic management deployments require a well-planned, interdisciplinary collaboration of design with operations and enforcement. Successful implementation also requires the following:
  - High-level champions who lead a culture change in an agency and institutionalize the agency's commitment to prioritizing traffic management
  - Overcoming the "we never did this before" attitude
  - Funding commitments for adequate long-term operational maintenance

England: roadside sign in Birmingham

England: sign on overhead gantry in Birmingham

Netherlands: roadside sign

Figure 9. Automated enforcement signs.

- Advancing active traffic management in the United States will require evolution of long-standing design practices, collaboration of design and operations disciplines, and advances in techniques to communicate with motorists in real time.

Findings for Design

- **Functionality of shoulders.** Representatives of the host highway agencies shared their evolving perspectives on the functionality of freeway shoulders. In both England and the Netherlands, it was noted that the need for the outside shoulder to serve as a disabled vehicle area has diminished because of improvements in vehicle mechanical reliability. Therefore, the risk level for not providing full shoulder widths may have diminished since fundamental freeway design criteria were first established. These types of considerations weigh into the host highway agencies' assessment of the tradeoffs for continual or dynamic shoulder running. Each of the countries visited had a general practice of reducing the speed limits in freeway sections where shoulder width was reduced (both permanently and part time) to allow shoulder running.
- **Shoulder running (or plus lanes) with variable speed limits.** On some motorway segments in England, Germany, and the Netherlands, the shoulder is used dynamically to create an additional travel lane when conditions are appropriate. When the travel lane is added on the outside edge (e.g., right side for Germany and the Netherlands, left side for England), "hard shoulder running" is the term generally

used. When the additional lane is on the inside edge, "plus lane" is the term used. Gantries that include speed and lane control signs are provided in these sections and can show a green arrow when the lane is available for use and a red cross when it is closed. The signs can also show the appropriate speed limit for when shoulder running is allowed or the plus lane can be used. In Germany, when a paved shoulder is converted to a travel lane, a reduced speed limit of 120 kilometers per hour (km/h) (75 miles per hour (mi/h) is considered (from a normal speed limit of 130 to 150 km/h (81 to 93 mi/h)). If reallocation of the roadway for hard shoulder running reduces lane widths to less than 3.5 meters (m) (11.5 feet (ft)), a speed limit of 100 km/h (62 mi/h) is instituted. During shoulder running, the speed limit of the hard shoulder and the general travel lanes varies based on data from surveillance systems (loop detectors and/or cameras)

- **Lane width.** When an existing roadway cross section is reallocated to add a lane, existing lane widths may be narrowed to accommodate the new lane. In several locations, lane widths varied within the cross section, with narrower lanes typically on the inside (or the lane nearest the median). In some instances, no-passing restrictions were instituted for trucks to restrict them from the narrow inside lanes, harmonize speeds, and maintain lane control.
- **Shoulder running and ramp junctions.** Different approaches are considered for shoulder running through ramp junctions. In England, initial operations of shoulder running used only shoulder segments between ramps (i.e., the shoulder functioned as a lane gain or lane drop at each interchange). In 2009 England implemented a pilot allowing through junction running on the M42 motorway at certain locations to increase capacity at key bottlenecks.
- **Lighting needs with shoulder running.** Lighting for shoulder running sections has been a discussion topic in England and, over time, the Highways Agency has found that continuous lighting treatments are not highly essential. In Germany and the Netherlands, continuous lighting is considered beneficial.
- **Variable speed limits, line control, and speed harmonization.** Speed harmonization is introduced through the use of variable speed limits to improve traffic flow on freeway sections that experience recurrent congestion and protect vehicles at the back of congestion-or incident-related queues. The speed harmonization system detects changes in traffic speeds and volumes along a corridor, and an algorithm automatically reduces speeds based on real- time traffic conditions. To ensure respect for the variable speed limits, communicating the reason for the lower speed and enforcement is essential. Representatives of the European agencies used the phrase "trust equals compliance" on several occasions to indicate that the speed limit needs to be reasonable and the reason for lower speed needs to be clear.
- **Gantry and detector spacing.** The spacing between gantries that contain variable speed limit and line control signs and detectors that collect traffic data varies among counties. In Germany the national standard is 2.5 km (1.6 mi), but Hessen spaces its detectors at 1 km (0.6 mi) and gantries at 1 to 1.5 km (0.6 to 0.9 mi). It justifies the closer spacing to collect better traffic flow data, provide better alternate route information, and improve system management. Other countries use 600-m (0.37-mi) to 1,000-m (0.62-mi) spacing of gantries. For gantry spacing, the countries visited

stressed the importance of having a continuum of information with intervisibility of signs on successive gantries for the driver.

- **Emergency refuge areas.** When the shoulder was used as a travel lane—either part time or permanently—emergency refuge areas were added. The spacing of the refuge areas varied by facility and country.
- **Signs.** There is an ongoing debate on the best balance between static and variable message signs. One thought is that variable message signs provide better opportunity to communicate with the driver, such as the reason for speed limit changes or the presence of a queue or anticipated delay downstream. Some suggest that all signs should be dynamic signs, whether electronic or mechanical. However, variable message signs are more costly and require backup power systems to maintain continuous operation during a power failure.
- **Evolution in design philosophy: transition to a performance- or risk-based design approach.** Representatives from England, Germany, and the Netherlands all emphasized the need to use performance- and risk-based methods for making design choices. Historically, highway design criteria have been developed with a static roadway in mind. On a dynamically operated roadway, the needs and solutions may differ from those of a statically designed roadway. Performance-based design is an outcome- based, operationally focused design approach that considers the desired goals and objectives of the transportation facility and establishes project design criteria accordingly. England has developed a risk-based approach to innovative design practices, providing additional flexibility to design for safe operations.
- **Evolution of design criteria.** Countries continuously evaluate cost-saving approaches, including the tradeoffs of increasing the spacing between gantries, detectors, and emergency refuge areas. In England earlier implementations are now considered conservative and experience indicates that greater spacing may be appropriate.

Findings for Performance Measures

- **Key performance measures: travel time reliability and safety.** The key performance measures used in some European countries call for improving travel time reliability while improving or maintaining safety. The active traffic management strategies being implemented allow a wide range of options to improve or maintain safety while providing substantive mobility benefits.
- **Other performance measures: travel speed and congestion.** Average travel speeds for a roadway section have been used to quantify successful implementation of traffic management strategies, in addition to recognized and documented improvement in congestion. In Germany, the Congestion-Free Hessen 2015 initiative was started with the intent to ensure continual improvement of traffic flow. The vision of the initiative is that "mobility is one of the greatest issues for the future in Hessen. Both in economic and ecological terms, as well as with reference to social and cultural aspects, this task demands our full attention. Because for a transit state like Hessen at the heart of Germany and Europe, mobility and logistics are not only sustainable

economic factors but also synonyms for a modern and progressive society."[9] Hessen has experienced an 80 percent reduction in congestion,[10] but the initial large reduction in congestion duration was because of the completion of major road projects.

- **Public relations.** Education of drivers and stakeholders on managed motorway features is important for successful operations. Projects are driven by desired outcome, so understanding the overall goal and clearly and successfully communicating the goal to the public are critical. Experiences in Europe have identified radio and Web-based approaches as the best methods to reach the public. In some cases, the driver culture of the area may influence how the treatments are implemented and communicated to drivers.

Findings for Planning

- **Safety concerns.** Politicians, citizens, designers, and implementers in England, Germany, and the Netherlands had concerns similar to those expressed in the United States about potential or perceived reductions or changes in safety because of the application of some management strategies. The Highways Agency in England developed a hazard index to systematically evaluate potential driver safety risks and aid in its decision to implement strategies and design choices on managed motorways. The agency uses a risk-based approach for transitioning the shoulder from an emergency lane to a travel lane. Its research has indicated that the risk of eliminating shoulders (at least for part-time use) is minimal.
- **Evaluation of feasibility.** Before managed motorway treatments were implemented, extensive studies were conducted to determine a technique or strategy appropriate to the problem and the roadway geometry.
- **Stakeholders.** It is important to bring all stakeholders (enforcement, trucking, traveling public, agency, and government leadership) in at the early stages of the planning and design process. Emergency management was a key stakeholder group to educate and strategize on in several European countries.
- **Legislation and policy.** In England, Germany, and the Netherlands, national or state policy was a driving factor in implementation of managed motorway concepts. In 2003, the German state of Hessen initiated Congestion- Free Hessen 2015, which specifically identified future technologies, traffic management, and mobility services as tools to optimize traffic flow and increase safety. In England, longstanding public concern about the environmental cost of highway expansion drove the development of various reports and policy initiatives that emphasized sustainability in seeking solutions to roadway congestion.

Findings for Lessons Learned

- **Corridors in progression.** There is an evolutionary path in the appropriate design and operational strategies of individual freeway corridors. As traffic and congestion levels increase in the corridor, different approaches and management strategies should be considered to accommodate changing needs, risks, and appropriate tradeoffs.

- **Effective use of space.** Several European countries dynamically manage the freeway space available. For example, they may use the paved shoulder space for traffic movement during peak travel periods and as a typical shoulder during off-peak travel times.
- **Importance of collaborative design process.** Actively and effectively managing roadways requires coordination across disciplines, and collaboration among planning, operations, and design is imperative. In England the Highways Agency uses the operational regimes to determine design criteria rather than adhere strictly to design standards.
- **Operating costs.** Stable, consistent, and ongoing funding for operations and maintenance is a critical component of the managed motorway concept.
- **Capital costs.** The M42 in England was designed conservatively on spacing of gantries, emergency refuge areas, and ancillary equipment. After monitoring operations and results, the English are making incremental changes based on data that demonstrate they can maintain or improve flow and safety while increasing the spacing between gantries and refuge areas and reducing lighting to lower costs.
- **Complementary treatments.** Many applications are complementary. For example, line control (or variable speed limits) and shoulder running installations result in complementary and synergistic operations and benefits.
- **Benefits.** The countries visited report that managed motorways result in improved safety, reliability, and air quality benefits and can be provided at less cost than traditional capacity expansion.
- **Public perception.** The countries recognize that a proposed operational scheme will be successful only if the public perceives it to be successful (despite what data may say).
- **Procurement.** Construction methods are evolving as a result of the high degree of technology required for managed motorway concepts. England has used innovative construction methods and offsite locations to assemble managed motorway gantries, signs, and ancillary equipment and realized efficiencies in buying equipment.
- **Sign messages.** England, Germany, and the Netherlands have found that it is important to test new sign messages with users before implementation.

Next Steps

As evidenced in this report, the scan team believes that much can be gained in the United States by implementing several concepts and strategies observed during the scanning study. The next critical step is the implementation phase. Scan team members are communicating key findings, promoting implementation ideas, and advancing the adoption of key approaches and practices described in this report. The scan team is also seeking champions from transportation agencies and organizations to implement policies and practices using flexibility and innovation in designing freeways for improved safety and operational performance.

1. INTRODUCTION

The purpose of this scanning study was to examine innovative design practices and techniques used in other countries to improve the operational performance of congested freeway facilities without compromising safety. This scan builds on other scans that focused on congestion management and managed lane programs.

Background

The Federal Highway Administration's (FHWA) Office of International Programs leads and coordinates efforts to implement international programs and activities that meet the priorities of FHWA and the U.S. transportation community.[1] The International Highway Technology Scanning Program serves as a means to access innovative technologies and practices in other countries that could significantly improve highways and highway transportation services in the United States. The program enables innovations to be adapted and put into practice much more efficiently without spending scarce research funds to re-create advances already developed by other countries. Personal domestic and international networking, team dynamics, and the creation of domestic champions for promising foreign ideas are keystones of the scan process. Successful implementation in the United States of the world's best practices is the goal of the program.[2] The program is undertaken jointly with the American Association of State Highway and Transportation Officials (AASHTO) and the Transportation Research Board's (TRB) National Cooperative Highway Research Program (NCHRP).

Planning for this scanning study began in February 2010 with a desk scan that recommended England, Germany, and the Netherlands as the countries to visit, along with Spain as the host of the 4th International Symposium on Geometric Design. In June 2010 a team of 10 U.S. transportation professionals with expertise in planning, design, and operations of freeways visited the four countries. Appendix A provides contact information and biographies for the team members. Table 1 lists the locations visited. During the scan, the team also attended the 4th International Symposium on Geometric Design and a workshop on managed motorways. Key findings from the symposium are in Appendix B. Appendix C provides the workshop agenda.

Purpose

Continued growth in travel on congested urban freeway corridors exceeds the ability of agencies to provide sufficient solutions and alternatives based on traditional roadway expansion and improvement projects. High construction costs, constrained right-of-way, statutory restrictions, and environmental factors are pushing agencies to explore solutions such as active traffic management and managed lanes to maximize throughput under congested conditions and improve safety by reducing collisions and nonrecurring congestion.

Table 1. Hosted locations for the freeway geometric design scanning study

Countries Visited	Locations Visited
Spain	valencia
Germany	Frankfurt, Hessen, and Mainz, Rheinland-Pfalz
Netherlands	Delft and Utrecht
England	Birmingham

The purpose of the scanning study was to examine design practices and techniques used in other countries to improve the operational performance and safety of congested freeway facilities. Finding cost-effective options to mitigate recurrent and nonrecurrent congestion on urban freeway facilities is one of the most significant challenges State and regional transportation organizations face. Internationally, transport agencies are using geometric design treatments linked to operational strategies, such as reallocating the roadway cross section, to dynamically reduce congestion while improving or maintaining safety of freeways in congested urban freeway corridors.

Managed Motorways—A Definition

"Managed motorways" is a term used in Europe and Australia.[3] Managed motorways combine actively or dynamically managed operational regimes, specific infrastructure designs, and technology solutions. They use a range of traffic management measures to actively monitor the motorway and, based on the monitoring, dynamically control speeds, add capacity, and inform road users of conditions on the network to optimize traffic and safety performance.

Examples of these measures include hard shoulder running, variable mandatory speed limits, lane control signals, incident detection and response, and driver information using variable message signs.

Managed motorways provide a significant opportunity to improve the capacity of motorways (called freeways in the United States) without acquiring more land and building large-scale infrastructure projects. Managed motorway concepts introduce new and revised operational activities that place greater reliance on technology than previously.

Managed motorways increase journey reliability and throughput of a motorway by speed management, and they increase capacity by hard shoulder running. Other techniques can be used to reduce disruption from joining traffic (e.g., ramp metering) and to improve safety (e.g., monitoring, detection, and emergency refuge areas).

For terms used in connection with managed motorways, see the Glossary.

Scan Team Members

The 10 scan team members (see figure 10) represented Federal agencies, State departments of transportation (DOTs), metropolitan planning organizations, research agencies, and private firms:

Figure 10. Freeway geometric design practices for improved performance scan team (from left to right, first row, Liz Young, Charlie Howard, Kay Fitzpatrick, Brooke Struve, Jim Rosenow; second row, Bart Thrasher, Jeff Jones, Martin Knopp, Mark Doctor, Greg Laragan).

- **Jeffrey (Jeff) C. Jones** (AASHTO cochair), assistant chief engineer of design, Tennessee DOT
- **Martin C. Knopp** (FHWA cochair), division administrator, FHWA Florida and Puerto Rico Divisions
- **Kay Fitzpatrick** (report facilitator), senior research engineer, Texas Transportation Institute
- **Mark A. Doctor**, safety and design engineer, FHWA Resource Center
- **Charles (Charlie) E. Howard**, transportation planning director, Puget Sound Regional Council
- **Gregory (Greg) M. Laragan**, assistant chief engineer, operations, Idaho Transportation Department
- **James (Jim) A. Rosenow**, State geometrics engineer, Geometric Design Support Unit, Minnesota DOT
- **Brooke A. Struve**, design program manager, FHWA
- **Barton A. Thrasher**, assistant State location and design engineer, Virginia DOT
- **Elizabeth (Liz) G. Young**, project manager and senior supervising transportation planner, Parsons Brinckerhoff.

Travel Itinerary

During the 2-week study, the team visited representatives in four countries and attended TRB's 4th International Geometric Design Symposium. The itinerary was as follows:

- **June 1:** valencia, Spain, team meeting
- **June 2:** valencia, Spain, Geometric Design Managed Motorway Workshop
- **June 2–4:** valencia, Spain, TRB 4th International Geometric Design Symposium
- **June 4:** valencia, Spain, meeting with hosts
- **June 5:** travel day
- **June 6:** Frankfurt, Germany, team meeting

- **June 7:** Frankfurt and Mainz, Germany, meetings with hosts
- **June 8–9:** Delft and Utrecht, Netherlands, meetings with hosts
- **June 10–11:** Birmingham, England, meetings with hosts
- **June 12:** Birmingham, England, final team meeting
- **June 13:** return to United States

Host Delegations

During the study, the team members met with representatives from various national and regional transportation agencies in the host countries. A list of individuals the team met with and contact information are in Appendix D.

Questions

To help the host countries address the team's interests, a set of amplifying questions was provided to them several months before the trip. The questions, in Appendix E, were grouped in four topics:

- **Geometric design.** Questions on geometric design were targeted to geometric design practices used to optimize the performance of existing or future freeway capacity. Examples of geometric design practices include dynamically varying the allocated use of the roadway section, reducing lane widths, and providing reserved areas for vehicle refuge, enforcement, or incident response and recovery. Also reviewed were signing, pavement marking, traffic control, lighting, variable speed limits and lane control, queue warning, and other elements considered in the design.
- **Performance measures.** The scan team asked questions on performance measures and metrics to evaluate, monitor, and report on geometric design alternatives aimed at improving the performance and flexible use of freeway facilities (e.g., full-time capacity addition, part-time capacity addition). The team sought information on existing and proposed methods, procedures, tools, public outreach, and techniques used to assess safety and operational implications or to compare geometric design alternatives, active traffic management strategies, and innovative uses of the cross section of a freeway.
- **Planning.** The scan team asked questions on the planning components used to integrate geometric design practices for improving freeway performance. The team was interested in organizational impacts, such as the need to address competencies, processes, structures, or other resource and agency leadership priorities; public acceptance or controversy; and political leadership and involvement.
- **Benefits and lessons learned.** Information was sought on the benefits of and lessons learned from different types of geometric design practices used to optimize the performance and flexible use of existing or future expanded freeway capacity.

Report Format

The purpose of this report is to describe the techniques being implemented in Europe to improve operations on motorways, summarize the findings from the scanning study, suggest strategies that might be applicable to the United States, and recommend activities that might increase awareness and knowledge of the need to and means for implementing managed motorway concepts. Chapter 2 summarizes the visits to each country. Chapter 3 summarizes key findings, and Chapter 4 outlines the scan implementation plan. Appendix A provides contact information and biographies for the team members. Appendix B presents findings from the symposium the scan team attended. The agenda for the workshop before the symposium is in Appendix C. Appendix D and E provide host country contacts and amplifying questions.

2. European Agency Approaches to Managed Motorways

This chapter describes the findings from each site visit. It provides a context for each country's overall approach to managing its motorways.

Spain

Meeting and Presentations

On June 24, 2010, the scan team met with and heard presentations from the following from the Ministry of Public Works:

- Jose A. Hinojosa, director
- Vicente Ferrer Perez, civil engineer
- Jose Yuste Maizal, civil engineer

Design Overview

Spain has few cases in which right-of-way for expansion is a severe limitation. Normally, it uses the median to increase the number of lanes or the land beside the highway. Most high-capacity roads in Spain (autopistas and autovias) are under the authority of the General Roads Directorate (Dirección General de Carreteras) of the Ministry of Public Works, a department of the central government of Spain. The Ministry of Public Works focuses on construction, while other departments, such as Enforcement or Police, handle operations. There is some coordination between operations and public works, but no strong ties. In addition to the central government, Spain has 17 local autonomous communities with responsibility for many highways. The working relationships of the General Roads Directorate with urban and regional traffic operations and planning agencies are often minimal.

Case Study Examples

The meeting with and presentations by the Spanish officials focused on a series of case studies illustrating Spain's approach to freeway design:[4]

Case Study 1. Widening from Two to Three Lanes

Since 2000, typical freeway and highway construction has included a 9- to 10-meter (m) (30-to 33-foot (ft)) median that is preserved for future expansion. Before 2000, there was no set requirement for median widths or expansion preservation. The Spanish use median and shoulder width for expansion from two to three lanes. Lane widths of 3.5 m (11.5 ft) are required on the interregional freeways and highways, with no exceptions. In this particular instance, the widening resulted in a 1-m (3.3-ft) outside shoulder and 50-centimeter (cm) (1.6-ft) inside shoulder (or shy distance). Roadway expansions are permanent, and the Spanish have not yet implemented shoulder or median lane use on a congestion-related or temporary basis. In some instances, they have reduced speed limits on the roadway because of sight distance restrictions from lane expansions. However, in this case, they were able to maintain the 80-kilometer-per-hour (km/h) (50 mile-per-hour (mi/h)) speed limit. Emergency pulloffs are located about every 1 km (3,300 ft).

Case Study 2. Tunnel Widening from Three to Four Lanes on Madrid M40

The approach roadway and tunnel section on M40 was widened from three to four lanes using the outside shoulder width. Traffic entering the roadway before the tunnel is restricted to the outside lane only, and merging to the adjacent traffic lanes is restricted (figure 11[5]).

Case Study 3. Table-Stayed Bridge (Puente) between Spain and Portugal

On the international bridge between Spain and Portugal, the Spanish widened the approach roadway and bridge section from one to two lanes in each direction using the outside shoulder width. Because the bridge deck was too narrow for two standard-width lanes, the speed limit on the bridge deck was reduced to mitigate the reduced lane widths (100 to 70 km/h (62 to 43 mi/h)). Lane widths for the bridge section were 2.85 m (9.3 ft) inside and 3 m (9.8 ft) outside. In addition to the speed reductions, truck traffic was restricted to the outside lane.

Source: Google Earth[tm].

Figure 11. Spain: view of approach and exit from tunnel[5].

Case Study 4. Carril Adicional, Reversible Use of Ramales (Madrid A6)

The Spanish implement a single contraflow lane on a two-lane separated, four-lane carriageway for weekend and summer traffic. The reversible lane is for high-occupancy vehicle (HOV) traffic only with one grade-separated entrance point. Transition from two-lane single direction to contraflow operation is via manual transition using hand-placed traffic cones.

Case Study 5: Shoulder Use for a Single-Lane Ramp

The Spanish allow shoulder use on a single-lane ramp to facilitate exiting traffic.

Case Study 6: Variable Speed Limits in Barcelona

An over-lane speed and lane control signal system is in place in Barcelona. Overhead gantries are located about every 500 m (1,640 ft). Two algorithms are used, one for traffic congestion and control and the other for air quality mitigation. The traffic congestion and control algorithm is most typically used. The lane control signals can close a lane (using a red X) to move traffic out of the path of a crash. The Spanish have experienced a decrease in crashes with the variable speed limit system, but specific data were not available. Speeds are reduced in 10-km/h (6-mi/h) increments, and 40 km/h (25 mi/h) is the minimum speed. Automated enforcement is conducted using cameras and mailed tickets. The variable speed limit system was implemented by the regional traffic agency in Barcelona.

Germany

Meetings and Presentations

The scan team had meetings with the following:

- Dr. Justin Geistefeldt, director of strategic traffic section, Hessian Road and Traffic Authority
- Reiner Dolger, Helga Rottenau, and Klaus Noll of the Ministry for Economics, Transport, Agriculture, and viniculture in Rheinland-Pfalz
- Dr. Kerstin Lemke of the Federal Highway Research Institute (BASt) after her symposium presentation on hard shoulder running

The scan team met with representatives from two of Germany's 16 states (Hessen and Rheinland-Pfalz). Responsibilities at the state level include the following:

- Planning, designing, constructing, maintaining, and operating motorways
- Finance
- Police and rescue services
- Framework for traffic information, including radio broadcasting

Federal government responsibilities include the following:

- Financing motorways, including equipment (such as intelligent transportation system (ITS) infrastructure)
- Setting technical guidelines and rules for motorways (as part of federal roads)
- Research and development (through BASt or dedicated projects)

Lemke's presentation[6] and paper[7] provided background information on hard shoulder running and safety studies conducted by BASt.

Hessen

In Hessen, Geistefeldt provided information on Congestion-Free Hessen 2015, hard shoulder running, and line control.[8]

Congestion-Free Hessen 2015

The Congestion-Free Hessen 2015 initiative started in 2003 with the intent to ensure continual improvement in the traffic flow on Hessen's roads. The vision of the initiative is the following:

> Mobility is one of the greatest issues for the future in Hessen. Both in economic and ecological terms, as well as with reference to social and cultural aspects, this task demands our full attention. Because for a transit state like Hessen at the heart of Germany and Europe, mobility and logistics are not only sustainable economic factors but also synonyms for a modern and progressive society.[9]

The initiative recognized that mobility is a crucial economic factor and that an important part of a work zone is the need to manage traffic. Along with improvements in flow, the goals are safer roads and reduced pollution. The project has three focus areas: future technologies, traffic management, and mobility services.

Under traffic management, several programs are being implemented. The following is an overview of each program from the pamphlet "Congestion- Free Hessen 2015: A Success Story."[9]

- The Hessen traffic center includes more than 80 monitors and data from more than 3,000 induction loops.
- Temporary use of hard shoulders is allowed during peak times. The Hessen traffic center opens the shoulder to vehicles based on traffic demand. The shoulders are monitored by cameras. Positive effects include the following:
  - Congestion reduced 30 percent
  - Crashes caused by traffic jams reduced 25 percent[9]
- A network control system using variable direction signs is used to provide information about alternate routes. Since 2005 additional dynamic information boards have been used. These displays can provide diversion information along with the reason for the diversion and expected delay. The additional information can increase acceptance of the diversion recommendations. Another feature of the network control system is the ability to provide delay data on signs. A model the state developed uses speed data from pavement sensors spaced at 3 to 4 km or at 1 km when within a line control section. The rule of thumb is that drivers will willingly reroute if travel time exceeds 20 minutes.[9] The contact noted that others display length of queue, but drivers want to know minutes and the value on the sign can have a large impact on driver behavior.[8]

- Long-distance corridors reflect cross-border traffic management with other German states. In the event of disruption to long-distance corridors, previously agreed-on diversion strategies are activated. Test corridors in use include the following:
  - South: Munich to Frankfurt
  - West: Frankfurt to Cologne
  - North: Dortmund to Hamburg
- Road works management involves scheduling road construction or maintenance to minimize the effect on traffic. Factors used and considered include an extensive historical database of traffic patterns, effective traffic models, planned special occasions (e.g., large trade fairs, sport events), and timing of maintenance on parallel routes.

Hessen has experienced congestion reduction of 80 percent, as shown in figure 12.[10] The initial large reduction in congestion duration was because of the completion of major road projects.

Line Control in Hessen

Another traffic management technique used is line control, known in other areas as variable speed limits. It is not called variable speed limits in Hessen because the signs can display more than just speed limits. The speed limit is set to harmonize traffic flow at high volumes. It decreases the variance in speed and capacity. Line control is used on all major freeways in the Hessen area. It is also used with hard shoulder running. The detectors are spaced at 1 km and gantries are 1 to 1.5 km apart. National standards are 2.5 km, so Hessen must justify the shorter spacing.

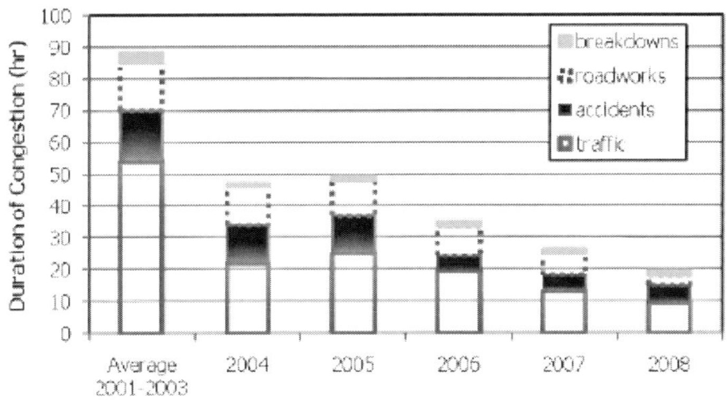

Figure 12. Germany: duration of congestion on motorways in Hessen[10].

Hard Shoulder Running in Hessen

The Hessen Web site[11] provides the following summary of the temporary use of hard shoulders:

> (Their use) for regular traffic increases the capacity of busy motorways at peak times, thus preventing traffic jams, or at least considerably reducing them. A high degree of safety is ensured by monitoring the traffic by video and by deploying systems to influence sections. Hard shoulders retain their intended function, apart from temporary use for

> traffic, which is to provide space to park vehicles in case of accidents, breakdowns, and during maintenance work. Using hard shoulders can temporarily increase capacity by up to 25 percent (given three regular lanes). Studies for the A5 motorway between the Frankfurt North-West interchange and Friedberg have shown that the benefits gained from avoiding lost travel time are so great that the system has paid for itself in less than 3 years. No serious impairments to road safety have been established to date. On the contrary: Studies for the A3 demonstrate that the higher capacity resulting from the use of hard shoulders noticeably reduces the potential congestion on a section, and thus the frequency of accidents caused by traffic jams. At present, hard shoulders are already in temporary use on 63 km (39.1 mi) in Hessen. Due to the very positive experience, further segments have been earmarked for temporary use of hard shoulders. Of these, around 12 km (7.4 mi) are presently undergoing implementation (A5 between the Darmstadt interchange and the Eberstadt junction in both directions).[11]

Hard shoulder running was first implemented in 2001. Frankfurt now operates 65 km (40.3 mi) of hard shoulder running. Hard shoulder control is usually integrated with line control systems. Both static and dynamic signs are used. The static signs have the arrows on a rotating drum that is changed depending on whether hard shoulder running is allowed (see figure 13). The dynamic signs show the following:

- Blank = hard shoulder running is not allowed
- Red X = disabled vehicle is on the shoulder
- Green downward arrow = hard shoulder running is permitted

Z 223.1 "Seitenstreifen befahren" (Use Hard Shoulder)

Z 223.2 "Seitenstreifen nicht mehr befahren" (End Hard Shoulder Use)

Z 223.3 "Seitenstreifen räumen" (Leave Hard Shoulder)

Figure 13. Germany: signs for hard shoulder running[7].

Figure 14 shows the dynamic impacts on congestion from temporary hard shoulder running on the A5.[8] Before 2001, more than 600 hours of congestion per year was experienced in the northbound direction. After hard shoulder running was implemented, congestion peaked at less than 200 hours of congestion. The southbound section did not experience as large a reduction, but most of the section did experience a reduction. The area near the interchange of Frankfurt did experience slightly more hours of congestion because of the transition to the nonmanaged section. Congestion, determined by sensors, was defined as speeds of less than 70 km/h (43 mi/h).

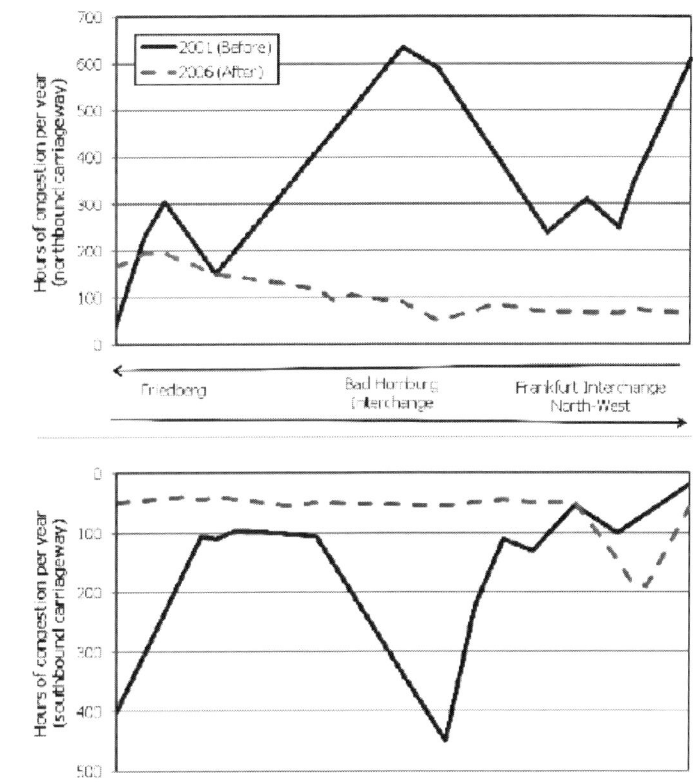

Figure 14. Germany: temporary hard shoulder running impact on congestion for A5[8].

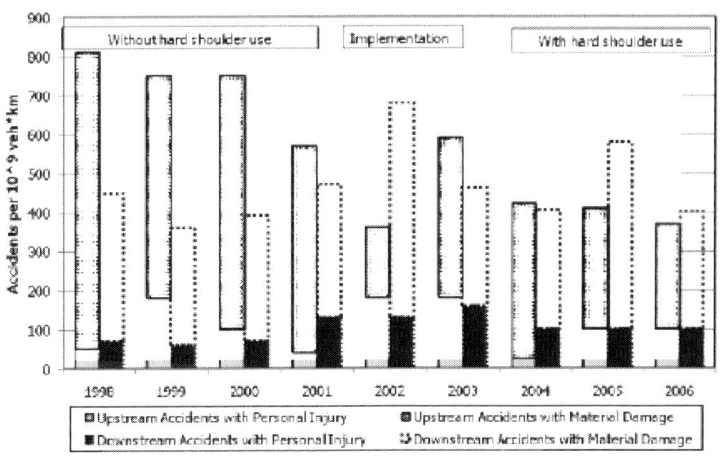

Figure 15. Germany: temporary hard shoulder running impact on safety for A5[8].

Figure 15 shows the impact of temporary hard shoulder running on road safety. The figure compares the years with hard shoulder use to the years during the provisional system (representing a 2-year implementation stage) and without hard shoulder use. The crash rate per hundred million vehicle-kilometers (10^9 veh*km) is similar or slightly lower for 2004 and 2005 compared to previous years. In 2005 crashes increased in the hard shoulder running section. Reasons for the increase are not known. The graph also shows a lower crash rate for the section of the motorway located upstream of the hard shoulder running, indicating that the safety benefits of implementing hard shoulder running extend beyond the location of the hard shoulder running. While benefits of hard shoulder running are apparent, Hessen takes a conservative approach and states that the system does not impact safety.[8] There were concerns that hard shoulder running would increase crashes, so finding no impact on safety is viewed as a positive result.

Based on the experiences on A5, the cost of the system was paid off in 2.5 years.[8]

General observations on hard shoulder running include the following:

- Acceptance for hard shoulder driving is high among truck drivers.
- Weaving maneuver challenges can occur at the end of the hard shoulder running.
- Trucks tend to move over to the hard shoulder once it opens.
- The Germans have not found any unique crash patterns around the hard shoulder running locations.

Hard should running is used only during congested periods; the shoulder operates as a shoulder when no congestion is experienced.

Long-Distance Corridors in Hessen

Since Hessen is a critical transportation hub for Germany and Europe, its traffic management and control systems must consider long-distance traffic from other German states. To be able to divert traffic in the event of major disruptions on the most important freeways, close cooperation between the neighboring German states is necessary. Cross-border traffic management on long- distance corridors was developed to facilitate this cooperation. In the event of a major traffic disruption on long-distance corridors, previously agreed-on diversion strategies are activated and the effectiveness of the freeway network beyond the state's borders is fully used. Initially this management strategy was successfully tested for the Frankfurt-Cologne corridor, and it has now been extended to additional corridors.

Rheinland-Pfalz

Rheinland-Pfalz has focused on the following traffic problems:

- Low accessibility in rural areas
- Recurrent congestion in urbanized areas
- Congestion caused by construction and severe weather
- Safety level (high in general, but problems remain on some roads)
- Lack of parking space for heavy goods vehicles (truck drivers are stopping on ramps when rest areas fill)

Lane control systems have been in operation since 1994 and rerouting variable message signs have been in place since 2009. Hard shoulder running was permitted after 2002 when the restriction on crossing the solid white line was removed.

Line Control Systems

Line control systems in Rheinland-Pfalz include the following:

- Dynamic speed limits
- Overtaking bans
- Warning signs

Currently, Rheinland-Pfalz has three line control systems in operation. It reports good experiences with capacity and safety and has found the systems particularly useful for heavy truck traffic and in mountainous areas with bad weather.[12] It considers overtaking bans important. It limits truck speeds because trucks are not permitted to overtake. Trucks are estimated to be up to 30 percent of the traffic on the autobahn.

Mobile police rather than automated speed enforcement is used. An estimated 80 percent of motorists comply with the speed limit. When speeds are not restricted by a speed limit, the 85th percentile speed on the motorways is estimated at 150 km/h (93 mi/h). Most drivers will accept the dynamic speed limit when the need is apparent. Dynamic speed limits have been set as low as 80 km/h (50 mi/h). At a location with cross slope concerns, different speed limits by lane have been used. For example, the right most lane was signed at 80 km/h (50 mi/h) and the other lanes at 100 or 130 km/h (62 or 81 mi/h).[12] Crashes influenced by a lane control system dropped 30 percent, while all crashes dropped 20 percent.[13]

Loop detectors and gantries are located every 2 to 3 km (1.2 to 1.9 mi). Many motorways were designed for 130 km/h (81 mi/h).

Dynamic Rerouting

Rhineland-Pfalz also has a dynamic rerouting system in place, providing drivers with alternate route information during congested, incident, or weather-related conditions. Even though the systems are not mandatory, they have experienced a high level of acceptance by drivers. Dynamic rerouting is particularly useful in urban areas and on dense roadway networks with available alternate routes.

Hard Shoulder Running

Hard shoulder running can be either temporary or continuous. The temporary systems require less planning and fewer environmental evaluations, so they can be implemented faster. Rhineland-Pfalz has one temporary system under construction. Officials provided the following observations on hard shoulder running to the scan team:[12]

- Noise level has been an issue, and officials have considered noise walls or a roof to encase the motorway to manage the noise.
- If they implement hard shoulder running, they try to remove any cross slope breaks.
- They have some permanent hard shoulder running, but not all have lane control.
- They believe safety has been good.
- In some locations, pavement reconstruction was required.
- Minimum lane width is 3.25 m.
- They have introduced speed limits (between 100 and 130 km/h (62 and 81 mi/h)) with hard shoulder running.

- Hard shoulder running sections have emergency pullouts about every 1 km. These sections are more difficult to operate and maintain. Temporary hard shoulder running sections must have cameras.

Overview of Hard Shoulder Running

According to Lemke,[7] the first pilot dynamic hard shoulder running scheme was implemented in 1996 on 1.6 km (1 mi) of the A4 freeway near Cologne. The section is activated in response to traffic volumes as a running lane, usually between 6 and 10 a.m. The section has one emergency refuge area. The speed is limited to 100 km/h (62 mi/h) during the entire day. Travel lanes were narrowed to 3.25 m (10.7 ft) and the shoulder width is 3.5 m (11.5 ft). The annual average daily traffic (AADT) exceeds 40,000 vehicles per day with almost 10 percent heavy traffic.

Based on the experience of pilot measures of permanent and dynamic hard shoulder running, the Federal Ministry of Transport, Building, and Housing defined a set of guidelines for the implementation of such measures. These guidelines, summarized from Lemke's paper, include the following:

- According to the German *Highway Capacity Manual,* it must be demonstrated that level of service defined as "D" could not be reached during peak hours.
- A width of at least 3.5 m (11.5 ft) must be provided for the heavy vehicle lane and at least 3.25 m (10.7 ft) for other lanes.
- Emergency refuge areas must be located at intervals of 1,000 m (0.62 mi).
- When hard shoulder running is intended to run through junctions, additional lanes must be provided in merge areas.
- A speed limit of 100 km/h (62 mi/h) must be established if restriping leads to lane width of less than 3.5 m (11.5 ft). This speed limit is also advisable in times without hard shoulder running.
- When paved shoulders are converted into travel lanes by restriping, a speed limit of 120 km/h (75 mi/h) should be considered.
- Prohibition of overtaking could allow for rescue services to pass through in case of emergencies.

The signs for use in hard shoulder running sections are shown in figure 13. Around 200 km (124 mi) of German freeways have dynamic hard shoulder running (see table 2). Although Germany successfully implemented measures of hard shoulder running, hard shoulders are still considered a vital element of freeway cross sections. Therefore, all cross section types of the new freeway design guidelines contain hard shoulders. The typical cross section for four-lane freeways (see figure 16), however, was designed so that hard shoulder running would be possible without restriping the roadway. The main design-relevant prerequisites of hard shoulder running were integrated into the design guidelines. The 0.75-m (2.5-ft) strips on both sides of the main carriageway shown in figure 16 are intended to stabilize the pavement and include the lane markings (shown as white squares in the figures). These strips are usually 0.5 m (1.6 ft) wide. On major motorways, their width is increased to 0.75 m (2.5 ft) to increase the sight distances in left curves and allow for more flexible planning of work zones (where lanes are shifted to the opposite carriageway).

Safety of Hard Shoulder Running

Lemke[7] reported on a study on road safety for hard shoulder running. Table 3 lists the sections included in the safety study along with their characteristics. Before-and-after data were examined for each section with hard should running along with comparison sites located upstream, downstream, or in the opposing direction. Comparisons were done for each freeway separately.

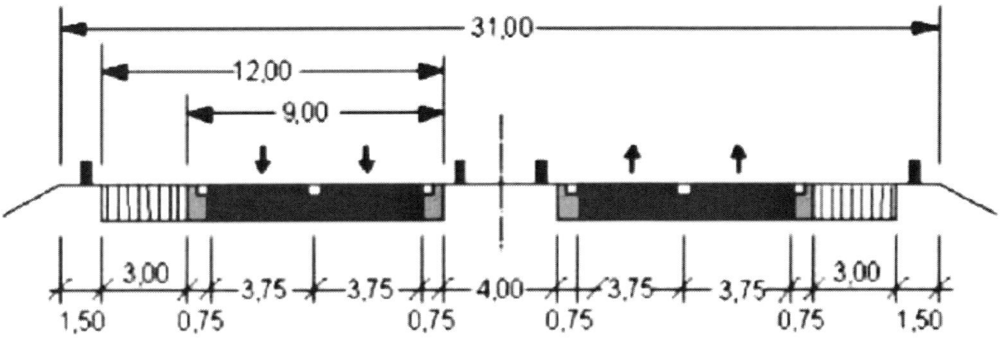

Figure 16. Germany: cross section for four-lane freeways allowing hard shoulder running[7].

Table 2. Germany: sections with dynamic hard shoulder running in 2009 [7]

Freeway	Section	Direction	Length (kilometer)	Lane no.
A3	Offenbacher Kreuz–AS Obertshausen	both	5.7 + 6.0	3 + 1
A3	Monchhofdreieck–AS Kelsterbach	both	3.3 + 2.0	3 + 1
A4	AS Refrath-AS Köln-Merheim	Köln	1.6	2 + 1
A5	AS Friedberg–Bad Homburger Kreuz	both	7.2 + 8.9	3 + 1
A5	Bad Homburger Kreuz–Nordwestkreuz Frankfurt	both	4.8 + 7.7	3 + 1
A7	Border HH–AS Kalten kirchen	Flensburg	22.5	2 + 1
A7	AS Neümnster-Süd–AD Bordesholm	Flensburg	14.0	2 + 1
A7	AS Soltau-Ost–Dreieck Walsrode	both	32.4 + 31.8	2 + 1
A8	AS Hofolding–AS Holzkirchen	Salzburg	9.8 + 9.8	3 + 1
A99	AK München Nord–AS Haar	both	18.0 + 18.0	3 + 1

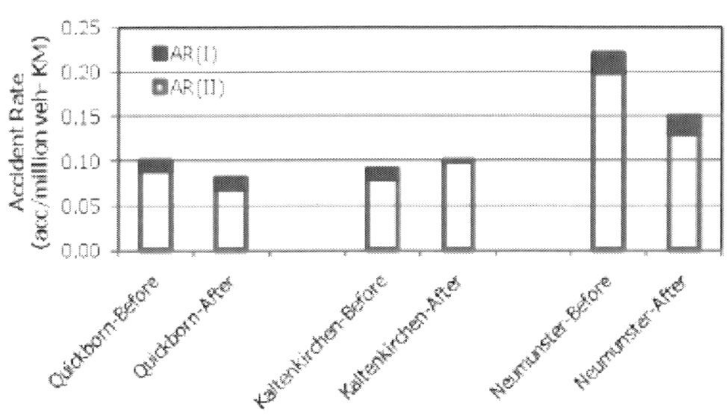

Figure 17. Germany: road safety on the A7 where AR(I) = fatal and severe injury crash rate and AR(II) = slight injury crash + severe property damage-only crash rate[6].

Table 3. Germany: sections included in safety study on hard shoulder running [6]

Freeway	Section	Type	Length in km (mi)	Width in m (ft)	Speed Limit in km/h (mi/h)
A4	Refrath	Traffic-activated	1.6 (1.0)	10 (32.8)	100 (62)
	Walldorf	Permanent	5.5 (3.4)	13 (42.6)	120 (75)
A6	Neckarsulm 1	Permanent	11.8 (7.3)	13 (42.6)	120 (75)
	Neckarsulm 2	Permanent	9 (5.6)	12 (39.3)	100 (62)
	Weinsberg	Permanent	8 (5.0)	12 (39.3)	100 (62)
	Quickborn	Traffic-activated	10 (6.2)	11.5 (37.7)	120 (75)
A7	Kaltenkirchen	Traffic-activated	12 (7.5)	11.5 (37.7)	120 (75)
	Neumunster	Traffic-activated	14 (8.7)	11.5 (37.7)	120 (75)

Figure 17 (see next page) shows one of the comparisons for the A7 freeway. The A7 freeway is characterized by commuter traffic to and from Hamburg as well as recreational traffic during summer holidays.

Hard shoulders on three sections going north (leaving Hamburg) are activated as running lanes in response to traffic volume. The 11.5-m (37.7-ft) cross section was restriped so that lanes had widths of 3.5, 3.8, and 3.6 m (11.5, 12.5, and 11.8 ft). Near the city of Neumunster, the section was only partly restriped so the hard shoulder is only 3.45 m (11.3 ft) wide, including the marking. The posted speed was set at 120 km/h (75 mi/h) during hard shoulder running.

The AADT is up to 35,000 vehicles. The portion of heavy vehicles is between 10 and 15 percent. The first two sites encountered while traveling north (Quickborn and Kaltenkirchen) had a crash rate below Germany's average crash rate (about 0.16 crashes per million vehicle-kilometers (0.26 crashes per million vehicle-miles)) before the implementation of hard shoulder running (see figure 17). For Neumunster, the safety level improved, bringing the crash rate to just below the country's average.

Lemke[7] concluded that well-designed hard shoulder running with a speed limit of 100 km/h (68 mi/h) when equipped with emergency refuge areas or a speed limit of 120 km/h (75 mi/h) when equipped with small (at least 2.5-m (8.2-ft) wide) shoulders can reach the same safety level as the average Germany freeway sections, which have the following rates:

- Fatal and severe injury crashes: 0.03 crashes per million vehicle-kilometers (0.05 crashes per million vehicle-miles)
- Slight injury crashes and severe property- damage-only crashes: 0.13 crashes per million vehicle-kilometers (0.21 crashes per million vehicle-miles).

An important prerequisite for this assumption is a detailed safety analysis of the section at the planning stage to eliminate high-risk road sites (especially near junctions). When the safety level is already better than these values before any hard shoulder running is implemented, no change is expected.

German Lane Widths

The German design manual, *Richtlinien fuer die Anlage von Autobahnen,* includes standard cross sections.[14] The standard cross sections for interurban autobahn (freeways) are shown in figure 18.

Figure 19 shows the standard cross section for four-lane highways, and figure 20 (see page 24) shows the standard cross section for intercity autobahns (freeways).

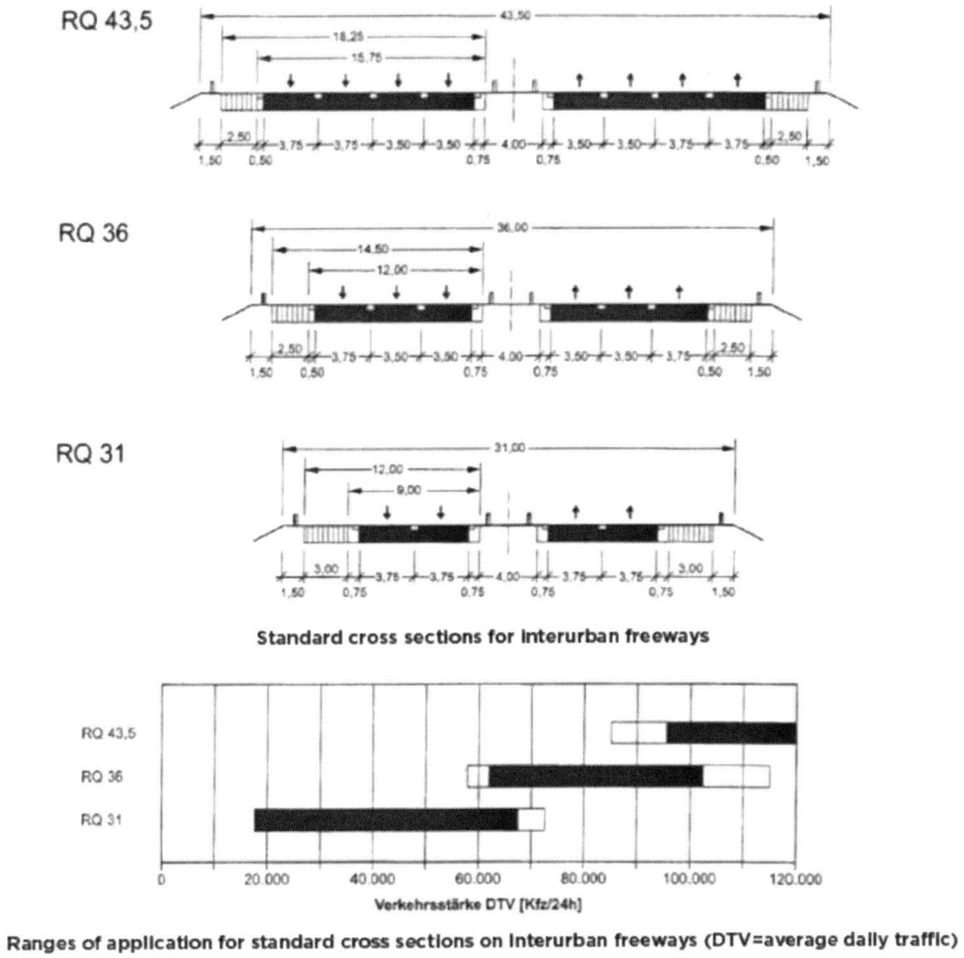

Figure 18. Germany: standard cross sections for interurban autobahn (freeway)[14].

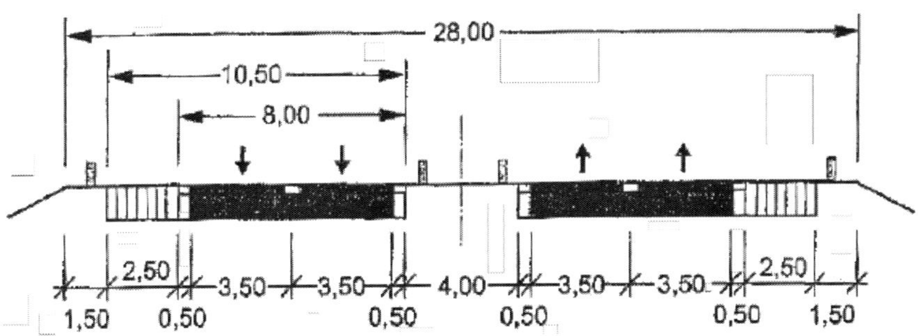

Figure 19. Germany: standard cross section for four-lane highways.[14].

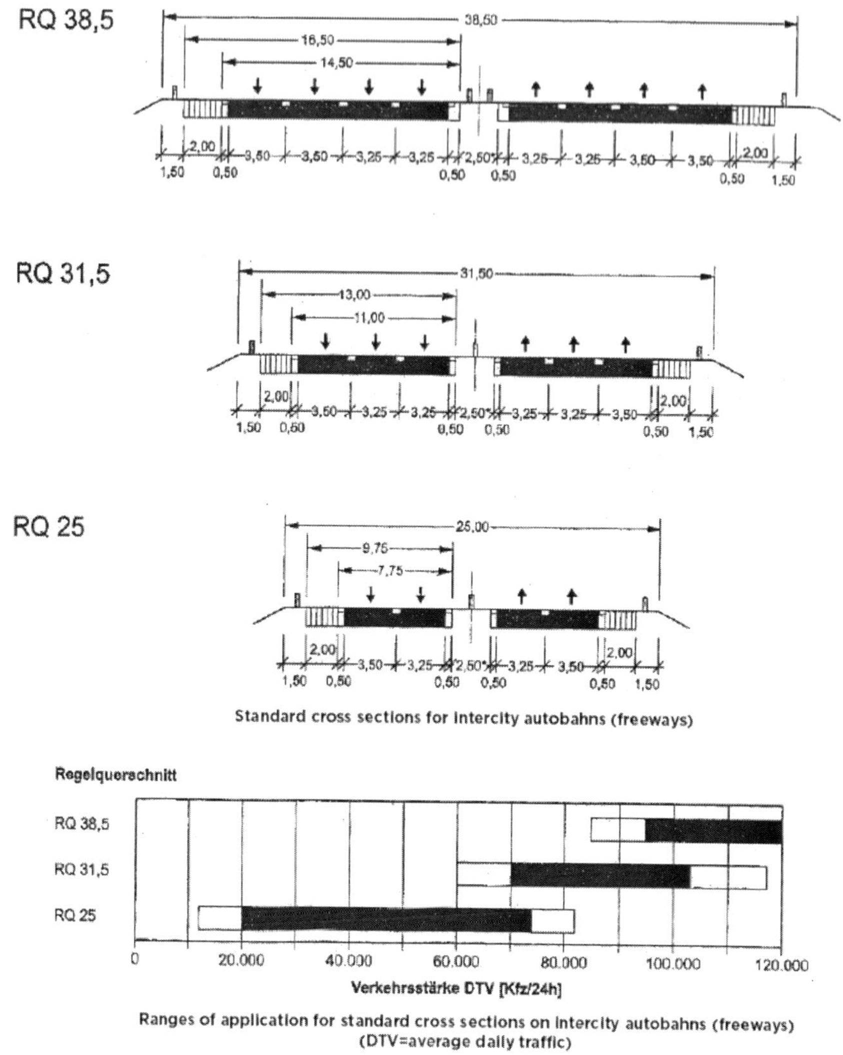

Figure 20. Germany: standard cross section for intercity autobahns (freeways)[14].

Left shoulder width is typically 0.5 to 0.75 m (1.6 to 2.5 ft). Previous design guidance called for a lane width of 3.75 m (12.3 ft). Current design guidance specifies lane widths of 3.5 m (11.5 ft) or 3.25 m (10.7 ft).

Netherlands

Meetings and Presentations

The scan team met with representatives of Rijkswaterstaat (Ministry of Transport, Public Works, and Water Management). The following made presentations to the scan team:

- Aad Wilmink, director of traffic management, provided the welcome.
- Alex van Loon discussed background information on the Netherlands to help the scan team understand the evolution of design in the country.

- Gerald Uittenbogerd presented on geometric design in the Netherlands.
- Bert Helleman discussed hard shoulder running and traffic management issues.
- Marthe van Dongen and Roel Nijsten discussed planning processes.
- Alex van Loon talked about future road plans and accompanied the team to view sites in Utrecht.

National Mobility Scheme

Rijkswaterstaat is responsible for the national road network in the Netherlands. Helleman provided the team with background information on transport policy and the managed motorway concept in the Netherlands.[15] Dutch transport policy has three pillars for national mobility:

- Limited extension of infrastructure
- Emphasis on better use of existing roads (Urgent Act on Road Widening)
- Road pricing

Dutch managed motorways have three key components:

- Hard shoulder running
- variable speed limits
- Interchange lane control

Hard Shoulder Running

The first hard shoulder running project was conducted in 1996. The use of the hard shoulder as an additional lane can occur two ways: hard shoulder running and plus lanes.

Table 4 lists current and planned road projects in the Netherlands.

In hard shoulder running, drivers can use the right shoulder as a travel lane. Refuge for vehicles is available at emergency refuge areas spaced every 1,000 m (0.62 mi). The speed limit is 100 km/h (62 mi/h) when hard shoulder running is permitted. The hard shoulder is used only during the peak period.

This approach results in less noise and fewer environmental concerns. Figure 21 shows an example of hard shoulder running in which the left shoulder is 1.1 m (3.6 ft). The figure also shows the German sign adopted by the Netherlands to communicate whether the shoulder lane is available. Figure 22 shows another example of hard shoulder running.

Table 4. Netherlands: road projects[15]

Time	Sections of Hard Shoulder Running and Plus Lanes in km (mi)	
	Hard Shoulder Running	Plus Lanes
Current	72 (45)	108 (67)
Planned (2010–2014)	151 (94)	90 (56)
Total	223 (139)	198 (123)

Figure 21. Netherlands: hard shoulder running[15].

Figure 22. Netherlands: hard shoulder running.

Figure 23. Netherlands: plus lane with former style of lane markings[15].

For plus lanes, the carriageway is restriped to reflect the permanent addition of another lane. Most of the additional lane width is from the left shoulder. The plus lane, the leftmost lane, is separated by typical lane lines and controlled by a lane control sign. The left lane is closed during off-peak times and the speed limit is higher. An advantage of the plus lane is at ramps (or junctions). The geometry is more conventional and familiar to drivers. The lane widths vary across the carriageway; the left lane is not as wide and has a width restriction (e.g., 2.3 m (7.5 ft)). Figure 23 shows an example of a plus lane. The photo, taken in 1998, does not include the pavement marking now used with plus lanes. Current markings are 9 m (29.5 ft) of stripe with a 3-m (9.8-ft) gap, commonly called 9-3. Figure 23 shows the typical lane marking of 3 m (9.8 ft) of stripes with a 9-m (29.5-ft) gap, called 3-9. Figure 24 shows plus lanes on A12 with current 9-3 lane markings and truck restriction signs. Figure 25 shows a closed plus lane.

Reasons hard shoulder running is used instead of permanently adding a lane include the following:

- Additional environmental concerns
- Noise
- Cost
- Need for more space for a permanent roadway expansion

Source: Google Earth.TH

Figure 24. Netherlands: A12 plus lane with current plus lane markings and truck restriction[5].

Source: Google Earthtm

Figure 25. Netherlands: A12 plus lane closed to traffic[5].

When hard shoulder running was not in use and the lane control signals were blank, between 6 and 60 violations occurred per hour.[15] In November 2006 the Netherlands introduced the use of a red cross to communicate that the hard shoulder is closed. Figure 26 shows the signs used when closing the hard shoulder. Figure 27 shows when hard shoulder running is not permitted, and figure 28 (see page 28) shows a close-up of a red cross indicating that hard shoulder running is not permitted.

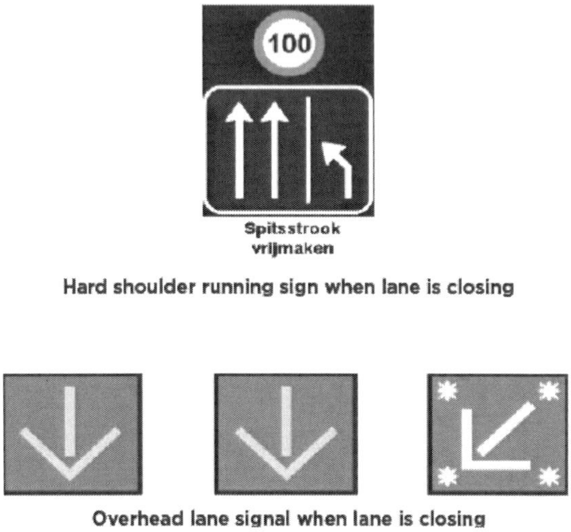

Figure 26. Netherlands: signs used to indicate hard shoulder running is closing.

Figure 27. Netherlands: example of when hard shoulder running is not permitted.

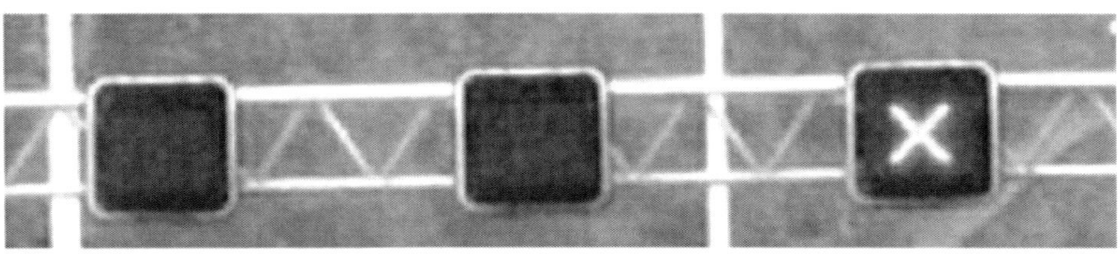

Figure 28. Netherlands: close-up of red cross used to indicate hard shoulder running is not permitted.

Gantries are spaced at about 600-m (1,970-ft) intervals. Hard shoulder running does occur through on- and off-ramp junctions. Pavement markings with different gap patterns are used in the merge area, as shown in figure 29 (see page 28) for off-ramp and figure 30 (see page 28) for on-ramp.

When an on-ramp is in a hard shoulder running section (figure 30), a solid white line indicates the location of the hard shoulder. The stripe is 20 cm (0.7 ft) wide and about 100 m (330 ft) long. It follows the same dimensions as gore striping.

Truck overtaking bans are not used when the hard shoulder running area is short so trucks are not forced to make multiple lane changes.

Figure 31 (see page 29) shows the sequence followed before opening hard shoulder running.

Safety measures employed with hard shoulder running include the following:

- Refuge havens (also known as emergency refuge areas)
- Automatic incident detection to warn operator of starting congestion
- Overhead signals and variable message signs
- Dynamic speed reduction (maximum speed limit is 100 km/h (62 mi/h))
- Roadway lighting
- Overtaking ban for trucks
- Incident management (maximum of 15 minutes for assistance to be on the scene)

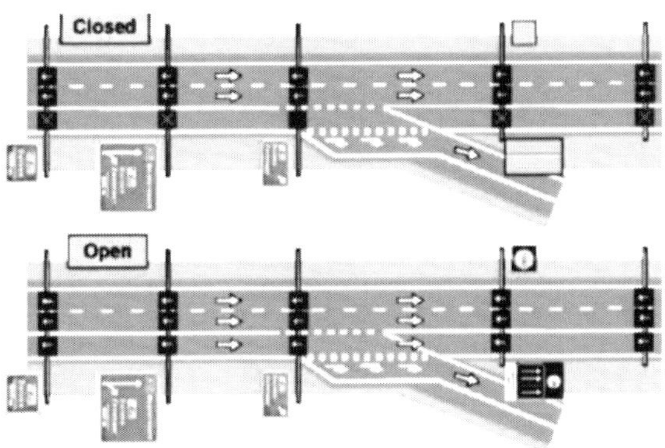

Figure 29. Netherlands: off-ramp design[15].

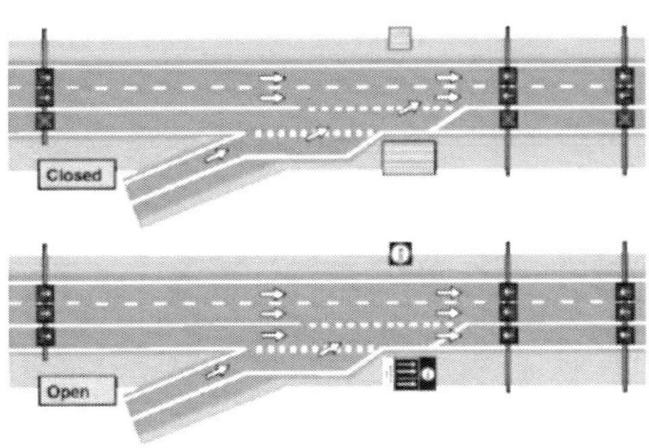

Figure 30. Netherlands: on-ramp design[15].

Figure 32 (see page 29) shows traffic safety data. It is based on 3 years of before data and 2 years of after data. Improvements were observed at most sites, primarily because of reduced congestion. Two hard shoulder running sites did have slight increases in crashes. The increases were attributed to a closely spaced ramp and exit design (weaving). The overall conclusion is that safety has been predominantly positive, caused by strong reductions in congestion. The Dutch have not observed an increase in hard shoulder running-related crashes, such as vehicles running on the hard shoulder colliding with a vehicle that has sought refuge on the shoulder. They are continuously looking for methods to detect incidents faster and minimize the number of incidents or disabled vehicles.

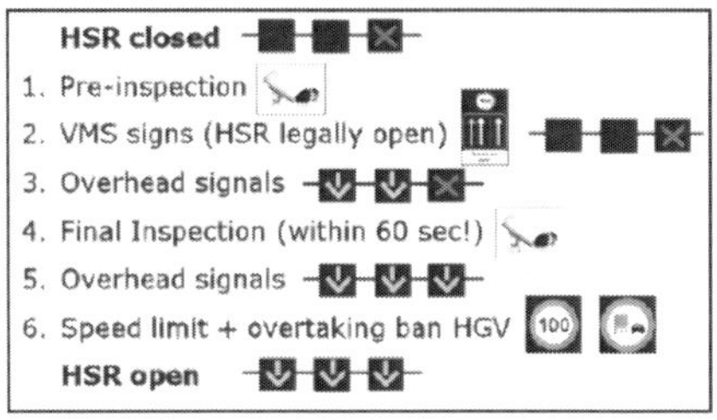

Figure 31. Netherlands: opening sequence for hard shoulder running[15].

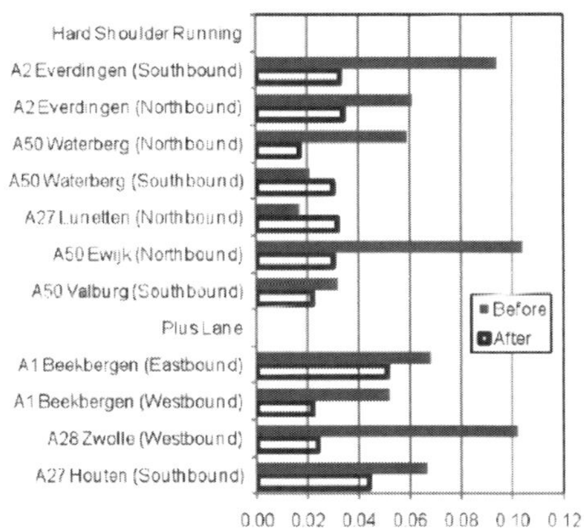

Figure 32. Netherlands: traffic safety at hard shoulder running and plus lane sites[15].

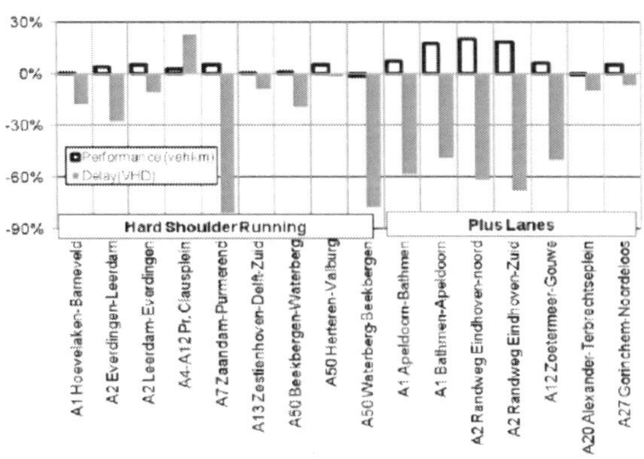

Figure 33. Netherlands: traffic benefits of hard shoulder running and plus lanes[15].

Figure 33 (see page 30) shows traffic flow benefits. Only one location experienced an increase in delay. For some sites, delay decreased over 90 percent. Larger decreases in delay were found for plus lanes than hard shoulder running. Performance showed increases for all sites except one.

Dynamic Speed Limits

The Dutch are examining speed limits through the Dynamic Speed Limits project, which was developed "to gain more insight into the impact of variable, tailor-made speed limits."[16] A pamphlet on dynamic speed limits provides the following discussion on the benefits:

Adjusting speed limits to unexpected and varying situations such as weather conditions, congestion, or an incident can improve traffic flow. It will also enhance the safety and improve local air quality. By applying a flexible and wide range of speed limits (50 to 120 km/h (31 to 75 mi/h)), operators can influence the actual situation on the road. Thus road users can drive faster if possible and need to slow down if necessary.

Tests are being conducted on several roadways and evaluations are ongoing. The evaluation on permanent speed reduction found excellent compliance (about 1 percent violations) because of enforcement. Reductions in emissions and noise level were also identified. Residents along the sections also expressed appreciation for the change.

Evaluation of variable speed limits is ongoing, and the Dutch are also investigating how to use data from traffic monitoring equipment to detect shockwaves and manage upstream speeds to limit the impact of the shockwave. A system called SPECIALIST (speed-controlling algorithm using shockwave theory) is being developed. Weather forecasts, such as a major rain storm, can also be considered when setting variable speed limits.

Interchange Lane Control

The host discussed an interchange junction where lane control is used to optimize the merging of two facilities. Figure 34 (see page 30) shows an aerial view of the interchange and a schematic of the merge area. The rightmost lane on the motorway is closed to facilitate the merging of two lanes from the ramp. Table 5 (see page 31) shows the results of the pilot study. Mean travel time was reduced and mean travel speed increased for both vehicles on the ramp (as expected) and vehicles on the main lanes of the motorway. Both approaches also experienced a decrease in vehicle- hours of delay, with ramp vehicles experiencing a slighter greater reduction.

Table 5. Netherlands: results of pilot study on interchange merge control[15]

Route Measure	Free Flow	Without Interchange Merge Control	With Interchange Merge Control	Percent Change
Red Route (Ramp)				
Mean travel time	4.76	11.03	10.42	-8 percent
Mean travel speed	98	41	45	+8 percent
vehicle-hours of delay		1,558	1,361	-13 percent
Blue Route (Main Lanes)				
Mean travel time	2.78	7.07	6.56	-7 percent
Mean travel speed	106	42	45	+7 percent
vehicle-hours of delay		1,455	1,398	-4 percent

Geometric Design

The Netherlands has dealt with challenges similar to those in the United States. For example, after the postwar rebuilding period came a period in the 1970s of "not in my backyard" opposition to building roads in certain areas. Between 1978 and 2004, the ministry conducted planning, design, finance, and maintenance tasks. In 2004, a major reorganization resulted in contractors doing the design work along with construction and maintenance. The ministry now focuses on planning, finance, specification, and audits. The motivation for the change was so that Dutch consultants could develop skills they could export to other countries. They use a design speed of 120 km/h (75 mi/h) with 3.5-m (11.5-ft) lane widths, 750-m (2,460-ft) minimum curve radius, and 5 percent maximum superelevation. They use spirals for transitions on any curves with radii of less than 2,000 m (1.24 mi). They use "Specific, Measurable, Achievable, Realistic, and Timely" (SMART) guidelines for geometric design. As part of this, they do not use any left-side exit ramps.

A philosophy of their design approach is to create self-explaining roadways with sustainable safety. They monitor research efforts in Germany, such as a large research project on the effects of superelevation on driving during heavy rain. The Netherlands also must manage its system during heavy rainfall. Rain results in pavements being wet about 12 percent of time. The Dutch have developed porous asphalt to minimize splash on the pavement surface.

Examples of features of current documents include the following:

- White (also known as black) spots, sight distance, distance between successive ramp terminals, design speed (now have 120-km/h (75-mi/h) design speed), transition superelevation, and others
- Special books for alignment, markings, etc.
- Use of AASHTO's *A Policy on Geometric Design of Highways and Streets,* also known as the *Green Book,* for reference

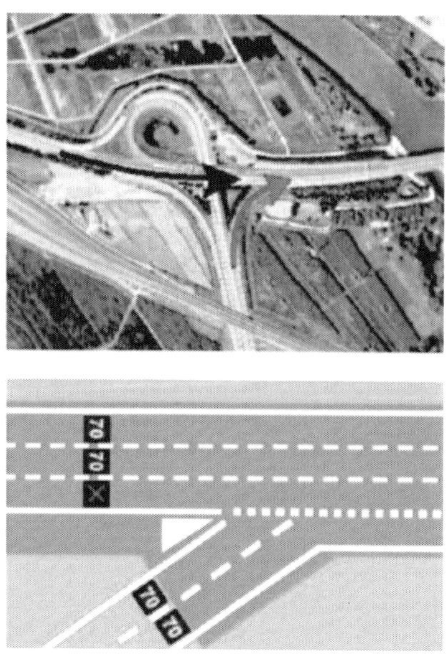

Figure 34. Netherlands: interchange with interchange merge control[15].

Table 6. Netherlands: cross section widths [17]

Vehicle Type	Function of *			Width in m (ft) for design speeds		
	Space of Interest	Space to left	Space to right	120 km/h (75 mi/h)	100 km/h (62 mi/h)	80 km/h (50 mi/h)
Passenger Car	Driving	Driving	Driving	3.35 (11.0)	3.10 (10.2)	2.80 (9.2)
			Emergency/ Recovery	3.25 (10.7)	3.05 (10.0)	2.75 (9.0)
		Emergency/ Recovery	Driving	3.25 (10.7)	3.05 (10.0)	2.75 (9.0)
			Emergency/ Recovery	3.30 (10.8)	3.10 (10.2)	2.80 (9.2)
	Emergency	Driving	No lane	3.15 (10.3)	3.15 (10.3)	3.15(10.3)
		No lane	Driving	n.a.	n.a.	n.a.
	Recovery Area	Driving	No lane	0.60 (2.0)	0.60 (2.0)	0.30 (1.0)
		No lane	Driving	0.60 (2.0)	0.60 (2.0)	0.30 (1.0)
Truck	Driving	Driving	Driving	3.35 (11.0)	3.35 (11.0)	3.35 (11.0)
			Emergency/ Recovery	3.30 (10.8)	3.30 (10.8)	3.30 (10.8)
		Emergency/ Recovery	Driving	3.30 (10.8)	3.30 (10.8)	3.30 (10.8)
			Emergency/ Recovery	3.35 (11.0)	3.35 (11.0)	3.20 (10.5)
	Emergency	Driving	No lane	3.15 (10.3)	3.15 (10.3)	3.15 (10.3)
		No lane	Driving	n.a.	n.a.	n.a.
	Recovery Area	Driving	No lane	0.60 (2.0)	0.60 (2.0)	0.30 (1.0)
		No lane	Driving	0.60 (2.0)	0.60 (2.0)	0.30 (1.0)

*Driving = driving lane Emergency = shoulder area or area for emergency use Recovery Area = between shoulder and barrier Emergency/Recovery = shoulder area or area for emergency use or area between shoulder and barrier.

The Dutch have debated lower design speed (politicians want lower design speed so roads take less space) versus higher design speed (drivers may be able to drive at 150 km/h (93 mi/h) in electric cars and not pollute the air).

International design law describes the minimum shoulder width of 2.5 m (8.2 ft), and a country must have good arguments to deviate from those standards.

Buses and trucks are 2.6 m (8.5 ft) wide, which is more or less the European standard. The Dutch use a design speed of 120 km/h (75 mi/h) for rural areas. In urban areas the design speed is 90 km/h (56 mi/h), but the definition of an urban area is not always clear. They believe they have impressive decreases in crashes in sections where speeds were reduced to 80 km/h (50 mi/h).

Lane width is 3.5 m (11.5 ft), but some believe that is too wide. Narrowed lanes provide space for hard shoulder running through redivision of the carriage way. In other words, it can provide greater capacity for a section. The Dutch design manual has a table that shows which lane to reduce given the conditions present.[17] Table 6 shows an example of the dimensions for the width of cross-section elements.

Traffic behavior is different when there are three lanes with no shoulder versus two lanes with a shoulder, so speeds are reduced when hard shoulder running is permitted.

Comments on horizontal curvature included the following:

- Minimum radius is 750 m (2,459 ft) for 100 km/h (62 mi/h), based on comfort, and 1,500 m (4,918 ft), based on visibility or sight.
- Research has shown a significant difference in crashes over and under 1,500 m (4,918 ft).
- Maximum superelevation is 5 percent.
- When using a 750-m (2,459-ft) radius, compensation must be made for the tight radius, such as by widening the carriageway.
- Under 2,000 m (6,557 ft), a spiral transition is used.

England

Meetings and Presentations

The scan team had meetings over a 2-day period with the following:

- Ruth Tilstone, Highways Agency
- Paul Johnson, Highways Agency
- Nick Hopcraft, Highways Agency
- Max Brown, Highways Agency
- Brian Barton, Highways Agency
- Andrew Page-Dove, Highways Agency
- Sarah Garland, Highways Agency
- Mike Wilson, Highways Agency
- David Grant, Highways Agency
- Paul Unwin, Highways Agency
- Lucy Wickham, technical dircctor, Mouchel

Highways Agency Strategic Plan

The Highways Agency is an executive agency of the Department for Transport responsible for operating, maintaining, and improving the strategic road network in England. Formed in 1994, the agency's aims are safe roads, reliable journeys, and informed travelers.

In *Strategic Plan 2010–15,* the Highways Agency's vision is "to be the world's leading road operator." [18] The strategic goals are as follows:

- We provide a service our customers can trust.
- We set the standard for delivery.
- We deliver sustainable solutions.
- Our roads are the safest in the world.
- Our network is a dynamic and resilient asset.

England faces constraints of space (land) and money and anticipates building few new roads. Officials are discussing how best to use existing assets, such as the hard shoulder, and how to minimize investment and maximize benefits. Wider challenges England faces include carbon emissions, financial constraints, and network resilience.

Controlled Motorways

The controlled motorway theory is based on relationships between speed and flow. Better highway flow can be achieved at lower speeds, which results in higher throughput. Therefore, the idea is to control the speed to maintain the flow. The slogan "go slower to get there faster!" communicates the idea.[19] Another benefit is improving journey time reliability. Improved reliability was emphasized in many of the Highways Agency's presentations. Using the controlled motorway concept results in fewer flow breakdowns and shockwaves, producing smoother journeys.

The controlled or managed motorway concept requires a different way of thinking about design and includes the following:[19]

- Operation of the road space is fundamentally altered by inclusion of the hard shoulder running concept.
- Operational regimes, rather than strict design standards, must drive the design of the scheme.
- The system must be operated and maintained.
- Understanding traffic flows and how a scheme will operate in all potential situations is needed.
- A new way of operating can introduce new hazards or reduce existing hazards. This requires a completely new approach to assessing the design of a managed motorway scheme to ensure safety is maintained.

Keys to success include creating an intuitive environment and anticipatory behavior along with driver compliance.

Managed Motorways

An outgrowth of the controlled motorway concept is managed motorways. Managed motorways stemmed from detailed desk studies and an onroad trial. A pilot scheme was implemented on the M42 in 2004, with hard shoulder running from 2006. Whether managed motorways should have a wider rollout after publication of results from the M42 was presented in the March 2008 *Advanced Motorway Signaling and Traffic Management Feasibility Study*.[20] Conclusions from the pilot and further analysis suggest the following about the managed motorways concept:

- Has high value
- Offers high journey time reliability benefits
- Can be delivered at 60 percent of the capital cost of conventional widening
- Shows safety benefits
- Shows environmental benefit [20]

M42 Pilot Study

In 2001, the secretary of state asked the Highways Agency to conduct a trial of hard shoulder running on the M42 between junctions 3A and 7 southeast of Birmingham (see figure 35), with the following objectives:

Figure 35. England: location of pilot test of hard shoulder running on the M42 in Birmingham[19].

- Optimize safety and performance, in accordance with the volume and makeup of the traffic.
- Provide more consistent journey times.
- Minimize harmful emissions and fuel consumption.
- Reduce delays and disruption from crashes and incidents.
- Provide improved warnings and traffic management in association with routine maintenance operations.[19]

Phased operation began in January 2005 and dynamic peak period hard shoulder running was implemented in September 2006.

The M42 pilot provided the opportunity to demonstrate and test a number of dynamic traffic management tools, positioned and used according to best practices, comprehensive safety hazard analysis, and predetermined operational regimes. The system included the following:

- Lightweight gantries with variable message signs about every 500 m (1,640 ft)
- Emergency refuge areas about every 500 m (1,640 ft)
- Appropriate road markings and fixed signing
- Continuous safety fencing
- Comprehensive closed-circuit television (CCTv) coverage with cameras typically at up to 250-m (820-ft) intervals
- Motorway Incident Detection and Automatic Signaling (MIDAS) loops
- Maximum 50-mi/h (80-km/h) speed limit (initially) for all lanes when the hard shoulder is in use
- Highways Agency Digital Enforcement Camera System (HADECS) for speed limit enforcement

- Lighting throughout the length of the scheme
- Necessary optical fiber cabling and communications links[19]

This infrastructure allows hard shoulders to be used as running lanes. Initially, this was implemented only during maintenance and incidents to prove that the concept was workable. It was followed by full-scale operation of peak-time hard shoulder running with the full involvement of emergency services.

The M42 active traffic management pilot accomplished the following:

- Development of a design tool to further develop controlled motorways and automatic enforced mandatory speed control activated by onroad loops and control algorithms
- Introduction of hard shoulder running and monitoring that has proved the validity of design and operation assumptions
- Monitoring of safety, environmental impact, and journey times, with results used to inform decisions on rollout of the managed motorways program
- Development of operation procedures to ensure safe use of the modified infrastructure (e.g., reduced barrier setbacks, narrow lanes with reduced sightlines that rely on driver compliance with speed and lane discipline).

Originally, the M42 pilot operated at a maximum of 50 mi/h (80.4 km/h) with the following results:[19]

- Average journey time has improved up to 24 percent in the worst p.m. peak time.
- Fewer vehicles experience speeds of less than 45 mi/h (72.4 km/h).
- Journey time variability during weekdays has dropped significantly, an average of 22 percent.
- On average, compliance is 94 percent or better at 70-mi/h (112.6-km/h), 60-mi/h (96.5-km/h), and 50-mi/h (80.4-km/h) speed limits.
- Personal injury crashes have declined from 5.08 per month before to 1.83 per month after implementation.

Following detailed safety review, the maximum speed limit was increased to 60 mi/h (96.5 km/h) with the following results:[19]

- Average journey times dropped (4 percent reduction).
- Average traffic speed increased (about 5 mi/h (8 km/h)).
- Speed limit compliance was maintained.
- Incidents did not increase; safety was maintained.
- The overall positive effect was that using a 60-mi/h (96.5-km/h) speed limit with hard shoulder running helped maintain speeds at higher levels during periods of congestion, giving a better overall performance than 50 mi/h (80.4 km/h) with hard shoulder running.

Compliance with Variable Mandatory Speed Limits

Compliance with the variable mandatory speed limit between January 2006 and May 2009 is shown in figure 36.[21] From late 2007 to early 2009, the mandatory speed limit was 50 mi/h (80.4 km/h) when hard shoulder running was permitted. During this time, compliance was not as good as when the mandatory speed limit was 60 mi/h (96.5 km/h) during hard should running (started in early 2009). Drivers did not see the reason for the lower speed and responded by driving at higher speeds. This behavior is also shown in the compliance when the mandatory speed limit was 40 mi/h (64.3 km/h).

The managed motorway concept introduced a new approach to operations and design. Both had to be considered together, along with overall traffic flows and the driver environment. In January 2009, the Highways Agency announced a £6 billion investment in the trunk road network. An additional 520 lane- miles (836.8 kilometer-miles) of capacity will be added, of which 340 lane-miles (547.1 kilometer- miles) will be managed motorways or hard shoulder running. The managed motorway program puts a stronger emphasis on the agency's role to actively manage the motorway network using technology in control centers and on the network. The use of technology is key to improving the reliability of the strategic road network. The managed motorway (or active traffic management) concept provides a toolbox of integrated approaches to manage the motorway environment.

The success of active traffic management on the M42 resulted in the Highways Agency being asked to extend the concept to the motorways around Birmingham, known locally as the Birmingham Motorway Box. Phase 1 of the Birmingham Box rollout (M6, M42, and M40) opened to traffic in December 2009. Work on phase 2 (M6 J8-10a) started in 2009–10, and it was scheduled to open to traffic in 2010–11. A through junction running trial started in December 2009. Table 7 (see next page) lists the planned program.

Operating a managed motorway required a new delivery and procurement approach. Needed was seamless transition through development, construction, and commissioning into actively operating the network. During development of the managed motorway concept, briefing notes were produced on 31 topics, including the following:

- Scenario options
- Case studies
- Driver demographics
- Driver education
- Position of vehicle in the driving lane
- Public and media opinion
- Candidate sites
- Use of surrogates
- Cross section crash studies
- Through junction running
- Emergency refuge areas
- Speed
- Hard shoulder crashes
- Environmental impacts
- Use of the driver simulator

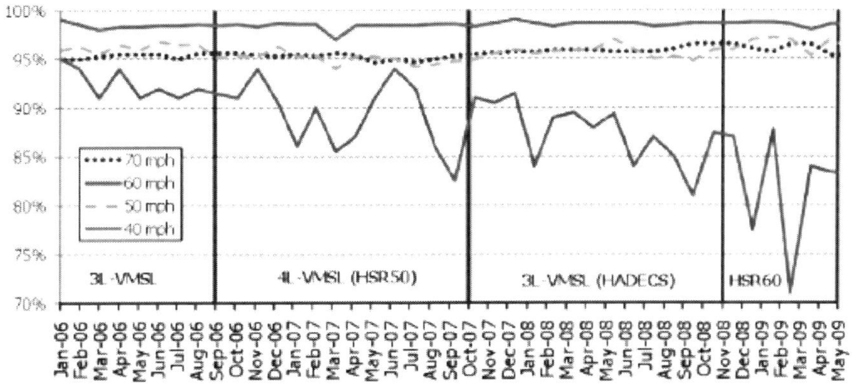

Figure 36. England: variable mandatory speed limit compliance.[21].

Table 7. England: Highways Agency managed motorway draft program[21]

Start Date	Scheme	Start Date	Scheme
2009–10	M4 J19-20/M5 J15-17	2015–16	M6 J2-4
	M1 J10-13		M6 J15-19
2010–11	M1 32-35a		M20 J3-5
	M6 J5-8 (B Box phase 3) M62 J25-30		M62 J10-12 M6 J13-15
	M62 J18-20	2016–17	M6 J21a-26
	M60 J8-12		
2012–13	M25 J23-27(DBFO s5)	2017–18	M1 J21-21a
	M25 J5-7(DBFO s2)		M60 J24-27 and J1-4
	M3 J2-4a		M1 J13-17
	M6 J10a-13	2018–19	M1 J17-19
	M1 J39-42		M5 J4a-6
	M1 J28-31		M1 J21a-23a
2013–14	M4 J3-21	2019–20	M6 J8 – M5 J2 (B Box 4)
2014–15	M27 J4-11		M23 J8-10
	M60 J12-18		M1 J35a-39
	M62 J26 M606 Link		M56 J6-8
		Post-2020	M5 J2-4a (B Box 5)

Web Sites

Interim Advice Notes (IANs), issued by the Highways Agency, contain specific guidance for use in connection with works on motorways and trunk roads in England, subject to implementation instructions contained in the IANs. Links to IANs are at www.standardsforhighways.co.uk/ians/ index. htm.

Recent IANs on managed motorways include the following:

- Interim Advice Note 112/08. *Managed Motorway Implementation Guidance—Through Junction Hard Shoulder Running.* June 2008.
- Interim Advice Note 121/09. *Advice Regarding the Application of Integrated Traffic Management.* June 2009.

- Interim Advice Note 111/09. *Managed Motorways Implementation Guidance— Hard Shoulder Running*. November 2009.
- Interim Advice Note 109/08. *Advice Regarding the Motorway Signal Mark 4*. April 2008.

General information on active traffic management is available at www.highways.gov.uk/knowledge/ 1334.aspx.

Managed Motorway Elements

Managed motorways refers to a toolbox of measures or techniques that include the following:[19]

- **Hard shoulder running:** using the hard shoulder dynamically as a running lane between junctions
- **Variable mandatory speed limits:** setting speed limits dynamically in response to congestion levels
- **Through junction running:** extending the dynamic use of the hard shoulder through the junction
- **Controlled all-lane running:** converting the hard shoulder permanently to a controlled running lane
- **Queue protection:** automatic queue detection and setting of lane signals and signs known as Message Sign Mark 4 (MS4s) to protect the back of a queue
- **Lane control:** setting Advanced Motorway Indicator lane signals (lane closed, lane diverted, reduced speed limits) to protect incidents and roadworks
- **Access control (also known as ramp metering):** regulating the amount of traffic entering the carriageway from the slip road to maintain the flow on the main carriageway
- **Driver information:** creating an informed driver environment by providing roadside information on variable message signs

Figure 37 shows components of the managed motorway. Supporting the managed motorway measures are a number of physical and operational features, including the following:

- **Monitoring.** The motorway is monitored through the use of CCTv and loop detectors (see figure 38).
- **CCTV cameras.** CCTv monitoring provides Regional Control Centres with real-time traffic flow and incident information. The cameras were spaced at 100-m (330-ft) intervals. The Highways Agency is now considering spacing of 200 m (660 ft) or more. New cameras will be low-light cameras to offset less use of continuous roadway lighting.
- **Loop detection.** Loop detectors are used to collect traffic data. The Highways Agency would like cameras to do more of the work of identifying incidents. Current camera technology, however, results in too many false alarms. In the agency's

experience, high quality control during loop installation results in low failures and false alarms. The Highways Agency does not consider radar and view loops critical.

- **Motorway Incident Detection and Automatic Signaling.** MIDAS is used on much of the Highways Agency network. It measures vehicle speeds and duration over loops and identifies when traffic flows have exceeded a predetermined level. The automatic system sets suitable speed limits to keep traffic flows at a predetermined level.
- **Incident detection.** CCTvs are monitored from a dedicated control room. Comprehensive roadside detection technology using loops indicates the need for response. Traffic officers monitor, respond to, and manage traffic flow. Hard shoulder monitoring cameras are also used. An observation on incident detection is that on a managed motorway, operators constantly watch the roadway, which could result in faster response times after incidents than for similar roadways without surveillance (which leads to the observation that it may be safer to have a breakdown on a managed motorway).
- **Emergency refuge areas.** These areas provide refuge away from the live carriageway (see figure 39). They are located about every 800 m (2,600 ft). vehicles are detected on entry by operators using CCTv and/or detection technology (loops). Each emergency refuge area contains an emergency telephone (see figure 40) linked directly to a regional control center.

Figure 37. England: graphic showing managed motorway parts[19].

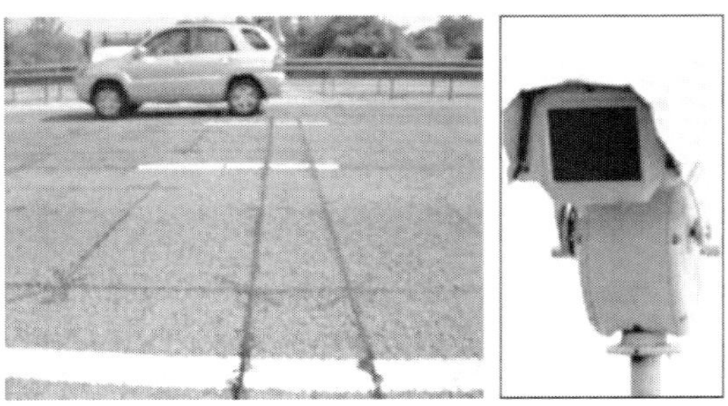

Figure 38. England: detection methods[19].

- **Gantries.** Gantries support lane-specific signals, the MS4 driver information panel, and signing (e.g., fixed direction or automated enforcement signs). Spacing for gantries is under review. The philosophy is that spacing needs to be sufficient so that drivers know they are still on a controlled roadway. A driver should lose sight of a gantry for no more than 1 to 2 seconds, based on Highways Agency advice. On the M42, gantry spacing is between 500 and 800 m. There is concern that compliance would be less if gantries were spaced too far apart. The ability to almost continuously see the variable speed limit provides reassurance to drivers that they are still on a controlled motorway. Gantries do not include a walkway. When walkways were present, there were problems with people climbing and walking on them. The walkways also did not benefit maintenance efforts. Eliminating the walkways resulted in a less expensive design.
- **Variable message signs.** variable message signs provide road users with onroad information both for the wider network area and locally. The signs provide text and picture information (see figures 41 and 42). They are fixed to lightweight gantries.

Figure 39. England: emergency refuge area and sign[19].

Figure 40. England: telephone in an emergency refuge area[19].

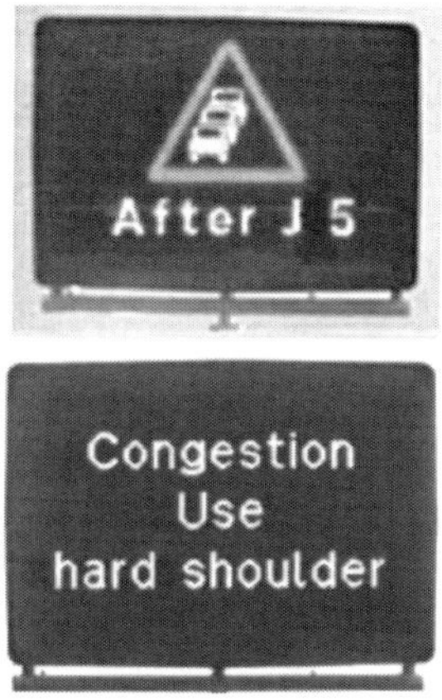

Figure 41. England: variable message signs [19].

Variable speed limit and shoulder control signs in use
(speeds are in miles per hour)

Closeup of variable speed limit sign

Sign being tested in shop

Figure 42. England: variable speed limit and shoulder control signs in Birmingham.

Figure 43. England: mandatory lane signals [19].

- **Mandatory signals.** Every usable lane has a signal indicator. The indicators depict speed limits (numbers) and driver instructions (arrows or Xs), as shown in figure 43 (see page 41). A red circle around the numbers indicates that the speed limit is enforceable. The signals are operated automatically by control system or manually by operators in the Regional Control Centre. When the signals are used to indicate an advisory speed limit, the beacons in the corner of the signal flash amber and the speed limit value does not have a red circle.
- **Speed limit.** The maximum national speed limit is 70 mi/h (112.6 km/h). It applies at all times (i.e., it is not posted on signs) unless the variable speed limit is active. The speed limit is reduced before allowing hard shoulder running. Mandatory speed limits are set automatically using traffic data from MIDAS loop detectors embedded in the road surface. Speed limits of 60 mi/h (96.5 km/h) and 50 mi/h (80.4 km/h) are displayed on overhead gantries to address congestion in response to the traffic conditions on the motorway. When necessary to protect traffic from queues, 40-mi/h (64.3-km/h) speed limits can also be set.[20]
- **Enforcement.** Automated enforcement is a component of the managed motorway system[19] The Highways Agency Digital Enforcement Camera System (HADECS) was "developed primarily to enforce the mandatory variable speed limits associated with schemes such as the M42 hard shoulder running and M25 controlled motorway schemes, whose safe and effective operation rely upon driver compliance with the displayed limits."[20] The system uses radar-based speed detection by cameras mounted on overhead gantries, linked to the lane signals displaying the speed limit. The agency uses a mix of cameras and dummies. When in operation, all evidence is automatically retrieved and recorded at a secured police office. Officials observed that it is an onerous process to conduct automated enforcement, including the part of the process involving the courts.[19] Compliance is important for managed motorways to function, and the agency conducts enforcement only to ensure compliance.
- **Roadway lighting.** The amount of roadway lighting needed is under debate.

Managed Motorway Safety

- The safety objectives for managed motorways are as follows:[22]
- Continue to contribute to the overall department's casualty reduction targets set by ministers.
- Use the procedures developed to assess road safety risk by detailed review of hazards and mitigations.
- Develop site-specific operational procedures that will ensure safe operation and maintenance of the modified highway.

Several methodologies are being used to ensure safety and demonstrate that the safety objectives are being met. A key component is the use of a generic program wide hazard log. Analyses are conducted of site-specific hazards to investigate whether they represent a new hazard or an increase or decrease of an existing hazard.

Another methodology is specification of design and operational requirements needed to control safety. Project hazard control review groups and a national safety group were established.

Understanding the risk profile of the network is important to maintaining safety. How the transition between operational regimes is handled is assessed. Relevant populations are assessed separately. These assessments are used to concentrate mitigations on key hazards. Steps used to verify the safety focus include the following:

- A safety report is generated.
- Accomplishment of safety objectives is demonstrated.
- Safety liabilities are understood.
- Whole-life ownership of safety requirements is defined.

Safety management activities include the following:

- Safety baseline and objectives
- Safety plan
- Risk assessment
- Hazard log
- verification
- Safety report
- validation
- Update of project safety documentation[22]

The safety plan, hazard log, and safety report must be logged before a project moves to the next stage.

Risks identified for typical motorway hazards include the following:

- vehicle stops in running lane.
- vehicle enters main carriageway unsafely.
- Tailgating occurs.
- Driver loses control of vehicle.
- Pedestrian runs in lane (live traffic).
- vehicle rejoins running lane.
- Motorcycle filters through traffic.
- vehicle reverses to exit slip.
- Individual vehicle drives too fast.
- vehicle stops on the hard shoulder when it is closed.
- Debris is in the running lane. [22]

Anticipated hazards from hard shoulder running include the following:[22]

- vehicles collide on hard shoulder while opening:
  - Switching between three- and four-lane running is excessive.
  - vehicle is stopped on hard shoulder as hard shoulder opens.

- Motorcycle is stopped on hard shoulder as hard shoulder opens.
- Crash occurs in or around emergency refuge area:
  - vehicle exits emergency refuge area during four-lane running.
  - Heavy-goods vehicle exits emergency refuge area during four-lane running.
  - vehicle is recovered from emergency refuge area during four-lane running.

Figure 44 summarizes early safety trends from the M42 pilot study. Observations from the evaluation include a decline in the personal injury crash rate of 50 percent, based on 3 years of data. Also, no crashes involving fatalities or serious injuries have occurred during hard shoulder running. Overall, incidents have dropped 25 to 30 percent.[22]

Managed Motorway Hazard Log

The use of a generic programwide hazard log is a key element of the safety program in England. Because the concept of a managed motorway is new and evolving, sufficient data are not available for a crash-based evaluation. A risk-assessment methodology was developed to guide decisions. For managed motorways, this took the form of a generic hazard log.

To develop the hazard log, more than 140 hazards were identified and scores were applied. The development of the risk level estimates required an understanding of how the motorway works. A thorough understanding of the interaction between operational schemes and design was needed.[23]

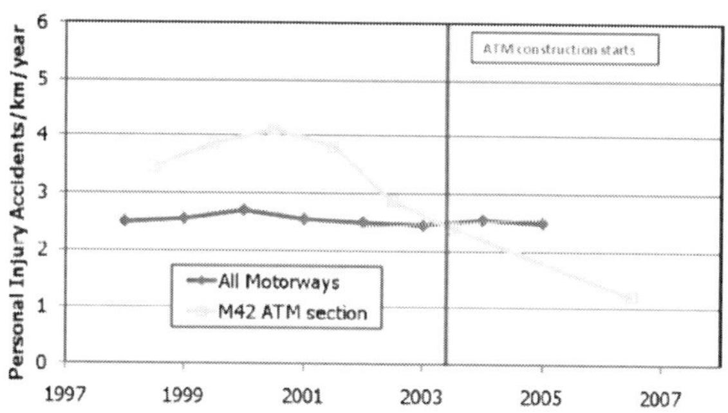

Figure 44. England: M42 pilot safety case study [22].

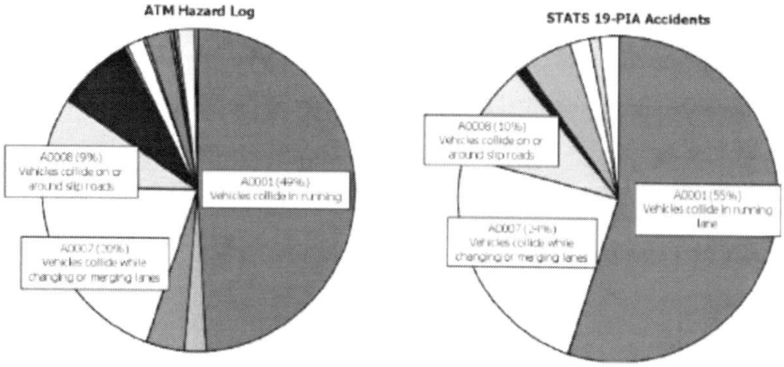

Figure 45. England: comparison of hazards for active traffic management site and other comparable sites[22].

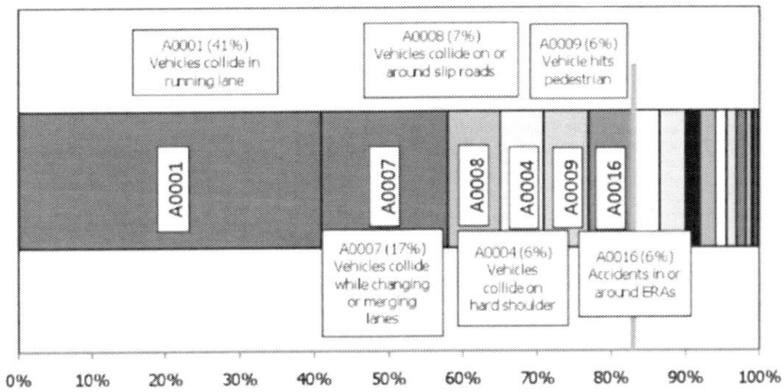

Figure 46. England: hazard log bar chart.[22].

The hazard log is prepopulated with the incidents, hazards, and causes known to be associated with such schemes. It is also prepopulated with the probability that a hazard leads to a crash and the severity of crashes. At the project level, the specific responsibility is to complete the entry of required information in the hazard log, including the project-specific frequency of different hazards.

Figure 45 shows a comparison of the types of crashes for the active traffic management site and other comparable sites.[22] The comparable sites use data from STATS 19, the form used by police in the United Kingdom to record personal injury crash data. The observation from the comparison is that hazard logs can be used to help manage the safety of new schemes and that findings for the active traffic management site compare well with STATS 19 sites. Note that active traffic management-specific crash categories, such as crashes in or around emergency refuge areas, were removed from the generic hazard log analyses to make the analysis more readily comparable to the STATS 19 data. Figure 46 shows the hazards in order of frequency for the active traffic management site. The most common hazard, representing 41 percent of hazards, is vehicles colliding in the running lane. The six most common hazards represent more than 80 percent of the total risk in the hazard log. The data shown in figure 46 include active traffic management-specific crash categories, such as crashes in and around emergency refuge areas.[22]

Reported Benefits

Sizable congestion and safety benefits have been realized with the active traffic management and managed motorway projects.[24] Figure 47 (see page 44) illustrates the change in average journey time profiles for the northbound direction between junctions 3A and 7. The time period for the before data was September to November 2005. The after period was for the same months in 2006. The change to four lanes with variable speed limits showed increased flows and reduced journey times. For example, the journey time for the Friday p.m. peak period dropped from more than 25 minutes to just above 15 minutes (see figure 47). Also reported were the following benefits:

- Journey reliability increased 22 percent when the hard shoulder operated at 50 mi/h (80.4 km/h).

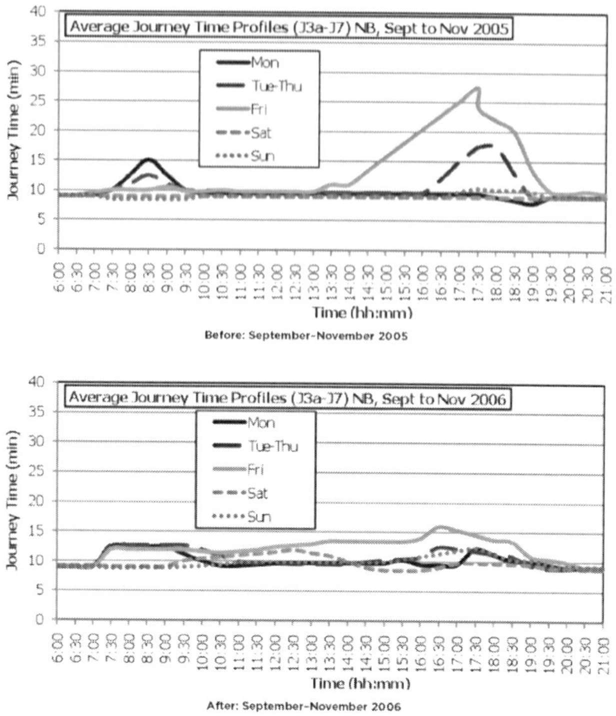

Figure 47. England: journey time improvements [24].

- Journey reliability increased 27 percent when the hard shoulder operated at 60 mi/h (96.5 km/h).
- Personal injury crashes dropped from 5.1 to 1.8 per month.[24]

Other reported benefits of managed motorways include the following:[21]

- Support the economy by providing capacity where needed
- Provide benefits at lower delivery costs than conventional widening (more capacity is delivered for the taxpayers' money)
- Result in less environmental impact than conventional widening
- Provide more reliable journeys for motorists
- Have a good safety record for the M42 managed motorway scheme
- Have been proven to emergency stakeholders as safe

Heavy vehicles are governed to 56 mi/h (90.1 km/h). Officials said that heavy vehicle operators like the ability of cars to accelerate and pass them legally when the speed limit is 60 mi/h (96.5 km/h) with hard shoulder running. They also like the improved trip reliability.[22]

Recent Developments

The rollout of managed motorways has prompted a number of modifications to the original design, most aimed at increasing efficiency or reducing the scope of the projects while ensuring that safety and economic objectives are met. Recent developments include the following:[22]

- Increase in maximum speed from 50 mi/h (80.4 km/h) to 60 mi/h (96.5 km/h)
- No road lighting of links
- Pilot site for through junction running on the M42
- Increase in nominal spacing of gantries and emergency refuge areas
- Improved hard shoulder monitoring
- Improved operating procedures

Another recent development is an examination of how best to communicate to drivers the availability (or restriction) of through junction running.[19] New legends are being developed and examples are shown in figures 48 and 49.

Figure 48. England: potential through shoulder running signs [24].

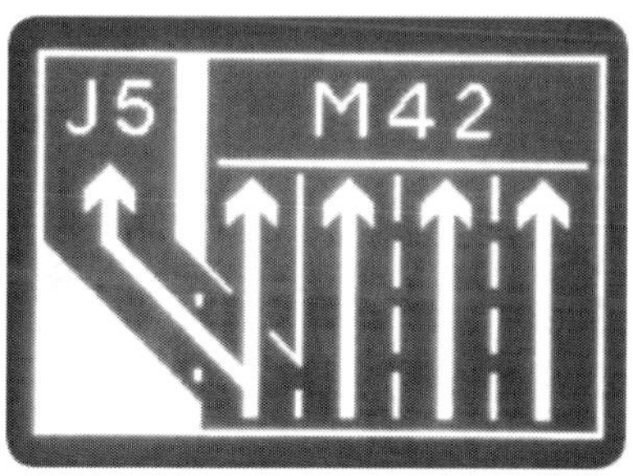

Figure 49. England: new legend for through shoulder running being tested[24].

Education

A key to success is the education of road users, including raising awareness, using the media, developing publications and Web sites, seeking buy-in from key stakeholders, and engaging emergency services. The results from a driver reaction survey included the following:[24]

- Ninety-three percent of participants who had used the hard shoulder believed the instructions for using it were clear.
- Eighty-four percent felt confident using the hard shoulder as a running lane.
- Thirteen percent of locals and 27 percent of long-distance users perceived it easier to join, change lanes, and exit the M42.
- Sixty-eight percent said that they felt more informed about traffic conditions.
- Sixty percent said it should be implemented elsewhere on the motorway network.

Advice provided to the scan team included the following:[24]

- Restrict interviews with the media until the day of launch.
- Include material on managed motorways in the training of new drivers.
- Focus on a Web site and radio (flash messages, 10-second sound bite, etc.). Have minimal use of leaflets (experience has shown low value and
- they cannot be used while driving anyway).
- Consider what highway codes may need to be altered (e.g., ability to cross solid white line).

Obtain early buy-in from key stakeholders, such as emergency services.

Traffic Officers

Before 2004, police officers managed traffic incidents on motorways. To provide faster response time, the Traffic Officers Services group was initiated.[25] This group is a part of the Highways Agency and is responsible for managing traffic incidents. In 2010 the Traffic Officers Services group had about 1,600 on staff, with about 1,000 on the road and 600 in control rooms. Normally, traffic officers operate only on motorways, but they operate on other roadway classifications during major events (e.g., visits of dignitaries). As a service agency, the group has performance goals such as travel time to an incident. Traffic officers have the power to direct traffic and close the motorway, but they do not have criminal authority.

The training program for traffic officers lasts about 3 years and includes 5-week foundation training and 9-month classroom room training along with several months of on-the-job experience. Their vehicles have amber and red lights rather than blue lights. Patrol routes are structured so officers can respond within their 20-minute response time goal. Phone calls are to be answered within 20 seconds, and a typical phone call takes 5 minutes.

3. KEY FINDINGS

Continued growth in travel on congested urban freeway corridors exceeds the ability of agencies to provide sufficient solutions and alternatives based on traditional roadway expansion and improvement projects. Several countries are implementing managed motorway concepts to improve the capacity of their freeways without acquiring more land and building large-scale infrastructure projects. Managed motorway concepts introduce new and revised operational activities that place greater reliance on technology than traditional roadway projects. Managed motorways combine actively or dynamically managed operational regimes, specific designs of infrastructure, and technology solutions. They use a range of traffic management measures to actively monitor the motorway and, based on the monitoring, dynamically control speeds, add capacity, and inform road users of conditions on the network to optimize traffic and safety performance. Examples of these measures include shoulder running, variable mandatory speed limits, lane control signals, dynamic rerouting, and the provision of driver information using variable message signs. Managed motorway concepts reduce collisions and improve journey reliability, travel time, and throughput on roadways through implementation of speed management and increase roadway capacity by allowing traffic to travel on the shoulder during peak periods. Other techniques can be used to reduce disruption from joining traffic (e.g., junction control and ramp metering) and to improve overall operations and safety (e.g., monitoring, detection, and emergency refuge areas).

Since a 2006 scanning study, active traffic management concepts have been implemented in Washington and Minnesota and are being considered in Virginia. During these implementations, several geometric design-related questions were voiced. In June 2010 a team of 10 U.S. transportation professionals with expertise in planning, design, and operation of freeways visited England, Germany, the Netherlands, and Spain. The purpose of the scanning study was to examine active traffic management design practices used in other countries to improve the operational performance of congested freeway facilities without compromising safety.

Key findings from the 2010 scan include the following:

- The European countries visited comprehensively integrate a suite of complementary techniques to dynamically manage traffic flow in response to changing volumes, speeds, and incidents. The result is demonstrably improved safety, travel time reliability, and congestion relief on urban motorway sections. These strategies and techniques are likely to provide great benefit if applied in the United States.
- The management strategies appropriate for a freeway corridor evolve as the needs and demands of the area change. In other words, transportation officials should recognize that freeways need a continuum of operational and management strategies that change as traffic needs and demands change.
- European countries have safety concerns similar to those in the United States and have successfully addressed those concerns in managed motorway deployments. Managed motorways have contributed to substantial safety improvements in Europe.
- Many European countries went through a paradigm shift in their design policies and practices by adopting risk- and performance- based approaches to making design choices on actively managed freeway facilities.
- Successful active traffic management deployments require a well-planned interdisciplinary collaboration of design with operations and enforcement. Successful implementation also needs the following:
    - High-level champions who lead a culture change in an agency and institutionalize the agency's commitment to prioritizing traffic management
    - Overcoming the "we never did this before" attitude
    - Funding commitments for adequate long-term operational maintenance

- Advancing active traffic management in the United States will require evolution of long-standing design practices, collaboration of design and operations disciplines, and advances in techniques to communicate with motorists in real time.

4. Implementation Strategy

The scan team used findings from its study to develop recommended implementation actions, which are summarized in this section.

- **Dissemination of information and promotion of findings.** The scan team believes that much can be gained in the United States by implementing several concepts and strategies observed and identified during the scanning study. To promote scan findings, team members have made technical presentations at several national meetings and conferences sponsored by FHWA, AASHTO, and other organizations to disseminate information from the scan. Team members also plan to make presentations to national and local meetings of the Institute of Transportation Engineers and other groups.
- **Research to support implementing recommended actions.** A research approach is needed to investigate the following issues:
  - Develop a synthesis of managed freeway geometric design practices as well as associated operational and safety performance for both European and U.S. applications. Identify the characteristics of circumstances in which managed freeway treatments could be successful.
  - Coordinate the design efforts with ongoing active traffic management research, synthesis, guidebook, and U.S. case study efforts underway by FHWA's Office of Transportation Management.
  - Conduct research on signing and marking needs in the United States for implementing shoulder running and assess the feasibility of implementing shoulder running in the United States.
  - Conduct research to assess if the hazard index approach developed by the U.K. Highways Agency may be directly or indirectly applied to U.S. conditions.
  - Conduct research to assess the air quality impacts of shoulder running and other managed motorway treatments.
  - varying lane widths in a cross section are being used in Europe. Conduct research to investigate applications in the United States and address key questions: How should the presence of a horizontal curve affect lane widths? What vehicle restrictions are needed for various lane widths?
  - Conduct a synthesis of practice to identify the best techniques to communicate with the public about the benefits and implementation of managed freeway treatments.
- **Research program for managed freeway concepts.** Also needed is a research program to investigate issues that are relevant to both managed freeways and all freeway design practices, such as the following:
  - Sign interpretation by the traveling public

- Driver understanding of symbols on roadway signs
- Stopping sight distance on horizontal curves
- Tort liability issues
- Dissemination of results from U.S. managed lane pilot projects

- **Design guidelines.** The following actions are suggested to assess the need to incorporate certain findings and recommendations of this scan into existing or proposed design guidance:
  - Query the AASHTO Technical Committee on Geometric Design for potential future additions to the *Green Book,* such as general language on managed lanes and emergency refuge areas, as well as potential enhancements to its treatment of flexible design practice.
  - Assess the AASHTO *Guide for High- Occupancy Vehicle (HOV) Facilities* for currency and consider broadening and updating it to include a managed freeway (active traffic management) focus.
  - Ask FHWA to clarify Federal policy and its position on the eligibility and use of managed freeway tools.
  - Query FHWA and the National Committee on Uniform Traffic Control Devices to determine possible updates needed to the *Manual on Uniform Traffic Control Devices* for managed freeway applications.

APPENDIX A. SCAN TEAM MEMBERS

Contact Information

Jeffrey (Jeff) C. Jones (AASHTO Cochair)
Assistant Chief Engineer of Design
Tennessee Department of Transportation
James K. Polk Building, Suite 700
505 Deaderick St.
Nashville, TN 37243-0349
Telephone: (615) 741-2831
Fax: (615) 741-0865
E-mail: jeff.c.jones@tn.gov

Martin C. Knopp *(FHWA Cochair)*
Division Administrator
Florida and Puerto Rico Divisions
Federal Highway Administration
Florida Division
545 John Knox Rd., Suite 200
Tallahassee, FL 32303
Telephone: (850) 942-9650, ext. 3019
Fax: (850) 942-9691
E-mail: martin.knopp@dot.gov

Puerto Rico Division
350 Avenue Carlos Chardon, Suite 210
San Juan, PR 0091 8-2148
Telephone: (787) 766-5600
Fax: (787) 766-5924

Kay Fitzpatrick *(Report Facilitator)*
Senior Research Engineer/Program Manager
Roadway Design Program
Texas Transportation Institute
Texas A&M University System
3135 TAMU
College Station, TX 77843-3135
Telephone: (979) 845-7321
Fax: (979) 845-6481
E-mail: k-fitzpatrick@tamu.edu

Mark A. Doctor
Safety and Design Engineer
Federal Highway Administration Resource Center
61 Forsyth St., Suite 17T26
Atlanta, GA 30303-3104
Telephone: (404) 562-3732
Fax: (404) 562-3700
E-mail: mark.doctor@fhwa.dot.gov

Charles (Charlie) E. Howard
Transportation Planning Director
Puget Sound Regional Council
1011 Western Ave., Suite 500
Seattle, WA 98104
Telephone: (206) 464-7122
Fax: (206) 587-4825
E-mail: choward@psrc.org

Gregory (Greg) M. Laragan
Assistant Chief Engineer, Operations
Idaho Transportation Department
PO Box 7129
Boise, ID 83707-1129
Telephone: (208) 334-8535
Fax: (208) 334-8595
E-mail: greg.laragan@itd.idaho.gov

James (Jim) A. Rosenow
State Geometrics Engineer
Geometric Design Support Unit
Minnesota Department of Transportation
Mail Stop 676
395 John Ireland Blvd.
St. Paul, MN 55155-1899
Telephone: (651) 366-4673
Fax: (651) 366-4699
E-mail: james.rosenow@state.mn.us

Brooke A. Struve
Design Program Manager
Federal Highway Administration Office of Infrastructure,
Program Administration
HIPA-20, Room E73-447
1200 New Jersey Ave., SE
Washington, DC 20590-9898
Telephone: (202) 366-1317
Fax: (202) 366-3988
E-mail: brooke.struve@dot.gov

Barton A. Thrasher
Assistant State Location and Design Engineer
Virginia Department of Transportation
1401 East Broad St.
Richmond, VA 23219
Telephone: (804) 786-5869
Fax: (804) 786-9311
E-mail: ba.thrasher@vdot.virginia.gov

Elizabeth (Liz) G. Young
Senior Supervising Transportation Planner
Certified Project Manager
Parsons Brinckerhoff
401 B St., Suite 1450
San Diego, CA 92101
Telephone: (619) 321-4181
Fax: (619) 338-8123
E-mail: youngl@pbworld.com

Biographies

Jeffrey (Jeff) C. Jones *(AASHTO cochair)* is the assistant chief engineer of design for the Tennessee Department of Transportation (TDOT) in Nashville, TN. Before that, he served as the director of the TDOT Design Division for 11 years. In that capacity, Jones was responsible for the roadway design criteria and plan development procedures used for projects in Tennessee. Before joining TDOT, Jones served as an engineer with the Federal Highway Administration (FHWA) for more than 10 years, including assignments in the Georgia and Tennessee Divisions. Jones has a bachelor's degree in civil engineering from the University of Tennessee at Knoxville. He is a licensed professional engineer in Tennessee. He serves on National Cooperative Highway Research Program (NCHRP) Technical Panels for Project 3-88 on Guidelines for Ramp and Interchange Spacing and Project 3-99 on Development and Application of Access Management Guidelines. He is a member of the American Association of State Highway and Transportation Officials (AAS HTO) Subcommittee on Design and Technical Committee on Geometric Design.

Martin C. Knopp *(FHWA cochair)* is the division administrator for FHWA's divisions in Florida, Puerto Rico, and the U.S. virgin Islands. As a key member of FHWA's management team, he exercises executive direction over one of the largest and most complex Federal-aid programs in the Nation. He leads a professional staff in providing guidance to the State of Florida and local officials in identifying surface transportation needs and implementing national transportation goals. He previously served as the FHWA Resource Center's Operations Technical Services Team leader. In this position, he led a team of traffic operations experts to provide technical assistance, training, and technology transfer services for program and project needs. Before rejoining FHWA in 2003, Knopp served for 4 years as the director of the Intelligent Transportation Systems (ITS) Division at the Utah Department of Transportation, where he led overall strategic direction of the ITS areas. He also chaired the ITS Subcommittee for the 2002 Winter Olympic Games in Salt Lake City and served as a consultant with the U.S. Department of State, assisting transportation and security organizers for the 2004 Summer Olympic Games in Athens, Greece. His FHWA assignments included traffic management engineer in the Georgia Division, traffic and safety engineer in the Utah Division, and ITS and Olympics operations engineer in the Utah Division. Knopp graduated from Pennsylvania State University in 1989 with a bachelor's degree in civil engineering. He is a licensed professional engineer in Utah.

Dr. Kay Fitzpatrick *(report facilitator)* is a senior research engineer at the Texas Transportation Institute. She has been the principal investigator for numerous research projects for the Texas Department of Transportation (TxDOT), NCH RP, Transit Cooperative Research Program, and FHWA. Fitzpatrick conducted the TxDOT research project that developed new criteria for design speeds over 80 miles per hour (128 kilometers per hour) and another TxDOT project that evaluated best practices on access to managed lanes in an existing freeway. She leads the FHWA project investigating engineering countermeasures for improving pedestrian and bicycle safety. She has bachelor's and master's degrees from Texas A&M University and a doctor of philosophy degree from Pennsylvania State University. She is a registered professional engineer in Texas and Pennsylvania. Fitzpatrick is a member of the Transportation Research Board (TRB) Operations and Preservation Group and the Operational Effects of Geometric Design Committee. She is a member of several Institute of Transportation Engineers (ITE) committees and wrote the chapter on intersection design for the ITE *Urban Geometric Design Handbook.*

Mark A. Doctor is a safety and geometric design engineer for FHWA's Resource Center in Atlanta, GA. He advances the use of flexible and innovative geometric design practices on a national level and within FHWA's design discipline. He also serves as the co team leader of FHWA's Design Focus Area Team on Interstate Access and Interchange Design and is responsible for coordinating the agency's internal communication and technical capacity-building activities related to interchange design and interstate access requests. Doctor has served with FHWA for more than 21 years in a variety of positions and geographic locations with responsibilities in project development, design, traffic engineering, and operations. He has a bachelor's degree in civil engineering from Clemson University and a master's degree in transportation engineering from the University of Florida. He is a licensed professional engineer in Georgia and serves on several TRB technical committees.

Charles (Charlie) E. Howard is the transportation planning director for the Puget Sound Regional Council in Seattle, WA. He is responsible for developing, updating, and implementing the long-range transportation plan for a metropolitan planning organization in a rapidly growing four-county region and directs the development of a regional congestion management process that is evaluating managed lane networks. Before joining the council, Howard worked for the Washington State Department of Transportation for 18 years, most recently as the director of strategic planning and programming in charge of statewide policy, planning, program development, and data functions. Howard is a graduate of Ohio State University and has a master's degree in city and regional planning from Harvard University. He chairs the TRB Transportation Systems Planning, Policy, and Process Section, is a member of the Future Strategic Highway Research Program capacity research development panel and the Association of Metropolitan Planning Organizations Policy Committee, and serves on several NCHRP research project panels on planning issues.

Gregory (Greg) M. Laragan is the assistant chief engineer, operations, for the Idaho Transportation Department (ITD) in Boise, ID. He manages the statewide construction, maintenance, materials, traffic, ITS, highway safety, and technical training activities of ITD. In the past, Laragan served as roadway design engineer, assistant district engineer, State traffic engineer, and assistant maintenance engineer for ITD. He has a bachelor's degree in civil engineering from Oregon State University and has been a registered professional engineer in Idaho since 1979. Laragan is a member of the Idaho Society of Professional Engineers, Intelligent Transportation Society of America, Women's Transportation Seminar, and ITE. He represents ITD on the AASHTO Subcommittee on Systems Operation and Management and chairs the panel for NCHRP Project 20-86 on Attracting, Recruiting, and Retaining Skilled Staff for Transportation System Operations and Management.

James (Jim) A. Rosenow is the State geometrics engineer for the Minnesota Department of Transportation (Mn/DOT) in St. Paul, MN. Rosenow leads the Geometric Design Support Unit, which is responsible for review and approval of preliminary engineering documents and geometric design exceptions on State highway projects. He is also a key person in the development of geometric design policy and criteria for Mn/DOT, and he directs and develops geometric design training programs for statewide audiences. Before that, Rosenow worked for 15 years in various highway design and project management capacities in the public and private sectors. He has a bachelor's degree in civil engineering from the University of Minnesota and is a registered professional engineer in Minnesota. Rosenow is a member of the AASHTO Technical Committee on Geometric Design (*Green Book* committee) and a friend of the TRB Committee on Geometric Design. He served as a member of the planning committee and as a small-group facilitator for Mn/DOT's Flexible Design for 21st Century Challenges forum in February 2009, and he served as a panel presenter at the Oregon Department of Transportation Practical Solutions Peer Exchange Workshop in July 2009.

Brooke A. Struve is a design program manager in the FHWA Office of Infrastructure. Her responsibilities include leading a team to develop tools and resources to advance expertise and practices on geometric design across the agency, participating on a similar team on the design of freeway interchanges, and reviewing complex proposals for access to the Interstate System. Before her current job, she was a project manager with the FHWA Office of Federal Lands Highway, leading a team designing highway construction projects in national parks, wildlife refuges, and forest highways. Before joining FHWA, Struve was a design team leader with the Utah Department of Transportation, where she led teams designing highway construction projects on freeways and urban arterials. Struve is a licensed professional engineer in Utah and is certified as a project management professional by the Project Management Institute. She is a member of the AASHTO Technical Committee on Geometric Design and has participated on several NCHRP panels.

Barton (Bart) A. Thrasher is the assistant State location and design engineer for the Virginia Department of Transportation in Richmond, VA. He is responsible for statewide oversight of all geometric and hydraulic design policies, standards, and technical guidelines. Thrasher has served with the Virginia DOT for more than 13 years and has more than 16 years of progressive and varied experience in the design and management of highway projects. He has a bachelor's degree and a master of business administration degree from Old Dominion University, as well as a bachelor's degree from Emory and Henry College. He is a licensed professional engineer in Virginia. He is a member of the AASHTO Highway Subcommittee on Design, AASHTO Technical Committee on Geometric Design, and TRB Committee on Geometric Design.

Elizabeth (Liz) G. Young is a senior supervising transportation planner for Parsons Brinckerhoff in San Diego, CA. Young works with FHWA to define the emerging practice of active traffic management (ATM) and develop a series of products that will guide ATM applications and practices in the United States. She works on a project team assessing the application of managed lanes, tolling, and ATM strategies for the Metropolitan Council, the metropolitan planning organization for the Minneapolis- St. Paul, MN, area. Young was the consultant project manager for the Washington State Department of Transportation's ATM Feasibility Study to assess the feasibility of implementing ATM strategies observed on the 2006 Planning and Designing Freeways for Congestion Management scanning study in the Puget Sound region. Young is a graduate of the University of Washington with a bachelor's degree in geography and a master's degree in transportation planning. She is an American Institute of Certified Planners certified planning professional and a member of the American Planning Association and Women's Transportation Seminar.

APPENDIX B. SYMPOSIUM FINDINGS

Overview

The scan team attended the 4th International Symposium on Highway Geometric Design in valencia, Spain, as well as a workshop on managed motorways held before the symposium. This appendix includes summaries of papers presented at the symposium and the workshop related to the issues examined by the scanning study. Also included are relevant findings from the desk scan.

4th International Symposium on Highway Geometric Design

The 4th International Symposium on Highway Geometric Design, held June 2 to 5, 2010, was sponsored by the Transportation Research Board. The symposium had 305 attendees representing 46 countries. The aim of the symposium was to stimulate continuous improvement in different areas of highway geometric design. It built on the success of symposiums held in Boston, MA, in 1995; Mainz, Germany, in 2000; and Chicago, IL, in 2005. They demonstrated that design practitioners, policymakers, and researchers worldwide face similar issues and can learn much by sharing their challenges and successes.

Workshop on Managed Motorways— Way Forward for the Future

John Smart of the Chartered Institution of Highways & Transportation and Jon Obenberger of the Federal Highway Administration coordinated the workshop on managed motorways. The presenters and their topics are in Appendix C. Smart and Obenberger provided the following introduction to the workshop:

With climate change on most countries' agendas, emission targets being imposed, financial constrains being imposed due to the current global financial situation, and the need to improve performance, there is a growing need to better manage and make better use of the highway assets that authorities control. Limited public funding, increasing construction costs, restricted urban environments, environmental constraints, and continued growth in travel are further limiting the ability of agencies to construct new capacity on congested urban motorways.

Looking to actively manage the cross section and operation of a freeway or motorway is now becoming commonplace in a number of countries. Managed motorways is a context-sensitive solution countries are pursuing to mitigate the detrimental effects of congestion within these corridors while optimizing the investment that has or will be made in the roadway infrastructure. Agencies are using innovative geometric design treatments in an attempt to optimize the utilization of the roadway cross section in support of actively managing traffic and dynamically using different operational strategies (e.g., speed harmonization, vehicle restrictions (occupancy, trucks), pricing or tolling) to improve the performance of specific lanes of the entire motorway.

The workshop will explore the issues technical, behavioral, and political surrounding the introduction of managed motorways on a strategic road network. The workshop will also examine the use of innovative geometric design techniques that are being used in support of these managed motorways.

Australia

Coordinated Ramp Metering

J. Cunningham reported on the Australian M1 motorway, which has coordinated ramp metering along the entire corridor.[26] The M1 connects to privately operated segments with tolls. The system functions only when needed with startup and shutdown of the ramp metering based on algorithms. When the motorway is saturated and the ramp metering is functioning, the system allows only as many vehicles on the motorway as are leaving it. The Australians have developed the *VicRoads Freeway Ramp Signals Handbook* and technical specifications for ramp metering devices. In Cunningham's words, unmanaged lanes are a chaotic system.

Sight Distance Around Concrete Barriers and Structures on Freeways and Interchanges

The provision of stopping sight distance around concrete barriers and structures on freeways and interchanges was examined by O.K. Arndt et.al.[27] They noted that strict application of the normal stopping sight distance model yields very wide shoulders that can act as parking areas and decrease barrier performance. Based on anecdotal evidence, ignoring sight distance requirements altogether is likely to decrease safety. Assuming a lower design speed can result in the design of inappropriate geomctric features that violate driver expectations and degrade road safety, especially if the design speed chosen is too low. The criteria proposed for sight distance around concrete barriers and structures in this paper were developed using less conservative, but realistic, values for many of the parameters in the stopping sight distance model. For example, the authors used a perception-reaction time of 1.5 seconds and noted that the value represents at least a mean value for surprise stopping for all drivers from several studies. Alternative deceleration rates and object heights were discussed, and higher deceleration and object heights than used in the current stopping sight distance procedure were considered in their sight distance calculations. Using these less conservative values was justified because the normal stopping condition that has been universally used is extremely conservative. The design stopping condition is a combination of several 85th percentile (or even higher) conditions. The use of 85th percentile conditions is common in road design. The combination of many 85th percentile values, however, can yield a very conservative value. The criteria developed by Arndt et al[27] retain the intents of the sight distance models by providing reasonable stopping and maneuvering capability. They noted that when using these criteria, it is paramount to not underestimate the design speed because of the smaller latitude in the models. Their new criteria comprise two sections:

- Sight distance over roadside safety barriers is possible.
- Sight distance over roadside barriers and structures is not possible.

The new criteria have been incorporated into the latest Australian road design guidelines (Austroads, 2009, *Guide to Road Design Series*).

Denmark

Ghost Drivers

In recent years, Denmark has seen a significant increase in the number of motorists driving against the traffic flow on motorways. The Danish name for these motorists is "ghost drivers." Figure 50 shows the 80 locations in Denmark where untraditional ghost drivers initiatives were established on or at motorway ramps, as reported by K. Kjemtrup.[28] The number of ghost drivers was 91 in 2006, compared to 65 in 2005. Kjemtrup noted that only about every sixth ghost driver was stopped by the police. Drivers were often under the influence of alcohol or confused (per police records), and the starting point of the ghost driver could not always be determined. Treatments used to minimize ghost drivers include the following.

- Traditional measures:
  - "Ghost gates" consisting of an extra set of "No Entry" road signs and two arrows marked on the ramp
  - Adjustment of the placement of road signs, the cutting of plants in front of signs, and restoration of road markings and arrows on motorway ramps
- Untraditional measures:
  - Special mandatory sign, "the Austrian hand," over a sign with the text "Turn Around" (see figure 51).
  - Flashing red lights beside the "No Entry" road sign (figure 51). When driving down the ramp, the road user activates the flashing red lights to the right of the road above the "No Entry" sign. According to Kjemtrup,[28] the system has been used for some years on seven ramps and has been established on 11 additional ramps.
  - Road lighting to make the access ramp more visible while the exit ramp remains in darkness. According to Kjemtrup,[28] the system is being tested at 14 selected access ramps.
  - Red flashing lane lights laid down in the road on exit ramps (figure 51). According to Kjemtrup,[28] the system has been established on a ramp and is planned for 10 ramps within 6 months.

Chevron Markings

According to P. Greibe, 45 percent of all injury crashes on Danish freeways are categorized as crashes with vehicles driving in the same direction.[29] Rear-end crashes typically occur in situations with very short headways and/or in combination with high speeds. Inspired by positive results from the use of chevron markings in the United Kingdom and France, a trial of chevron markings on Danish freeways was conducted in 2007. The markings were established on five road sections (each 4 kilometers (km) (2.5 miles (mi)) long) and consisted of a series of white arrowheads on the road surface at 36-meter (m) (118-

foot (ft)) intervals. The chevron markings were accompanied by roadside signs advising drivers to keep a distance of two chevron markings from the vehicle in front. Figure 52 (see next page) shows photos of the markings and the signs.

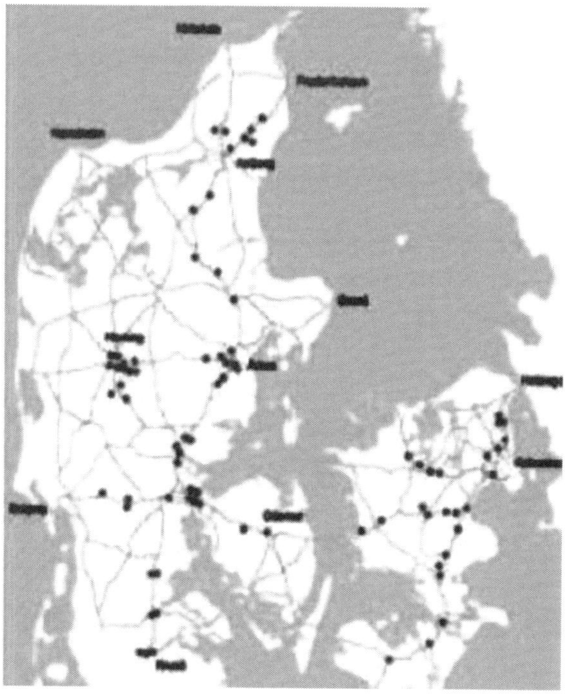

Figure 50. Denmark: locations with ghost driver initiatives[28].

Figure 51. Denmark: treatments for ghost drivers[28].

Two months after the markings were installed, significantly fewer vehicles had small gaps (less than 1 second). The largest reduction was in the left lane. The number of vehicles with gaps less than 2 seconds was also reduced, but the reduction was smaller. Speed was reduced slightly. Data were collected in 2010 to determine the long-term (2-year) effect. Telephone interviews were conducted 4 to 6 months after the chevron installations. Of the 916 respondents, 80 percent noticed the chevron markings. Of those 80 percent, 96 percent knew the purpose of the markings.[29]

Netherlands

Sustainable Safety

Rapid economic growth generated increased traffic demand in the 1980s and 1990s, which contributed to increased safety concerns on the Dutch roadway network. In 1998 the principles of sustainable safety were adopted in most of the design guidelines in the Netherlands. One method used to put the sustainable safety concept into practice was establishment of three classes of roads:

- Roads with a through function for rapid movement of through traffic
- Roads with a distributor function
- Roads with an access function

Each road class must comply with certain functional requirements, making each category different and readily recognizable to the road user. Pavement marking is an example of how the roads are to look different. As figure 53 shows, rural access roads have a driving lane with a broken edge marking, distributor roads have a broken edge marking plus a double continuous centerline, and regional through roads have a continuous edge marking and double continuous centerline with a green marking in between. About 40 percent of the rural distributor road network had been provided with the essential road marking strategy.[30]

Figure 52. Denmark: chevron markings[29].

Figure 53. Netherlands: pavement markings for three road classes [30].

During the past 10 years, the Netherlands has widely implemented the vision and ambitions of the sustainable safety program.[30] The entire road network has been reclassified and more than 70 percent of the urban network has been converted (many of these roads with some form of physical speed control measures) to 30-kilometer-per-hour (km/h) (19-mile-per-hour (mi/h)) zones and nearly 60 percent of the rural network to 60-km/h (37-mi/h) zones (about half of which have physical speed control measures). In addition, significant strides have been made to realize the essential recognizable characteristics, with more than 40 percent of the 80-km/h (50-mi/h) distributor roads and 70 percent of the 60-km/h (37-mi/h) access roads being provided markings that comply with the new guidelines. About 2,000 roundabouts have been built. These efforts have prevented an estimated 300 to 400 deaths per year.

Mobility Policy

A mobility policy was developed, focusing on three aspects: reducing congestion, increasing reliability, and shortening door-to-door travel times.[30] To achieve the goals, the policy relies on construction of new road space, road pricing, and better use of existing road space.

The current taxes on vehicle mass, fuel type, and new vehicles will be replaced with a charge per kilometer traveled on all roads. Therefore, motorists will pay road taxes based on how much they actually drive their vehicles instead of paying a flat rate for owning a vehicle. The proceeds from the kilometer charge will not exceed the combined proceeds from the old taxes. Motorists who drive relatively little will benefit from the system. The proceeds from the kilometer charge will go to a new infrastructure fund for financing investments in roads, railways, and related infrastructure.

Better use focuses on making the best possible use of available road capacity.[30] It is defined as the best possible management of traffic demand over a road infrastructure supply. Examples include extending a weaving section or entry or exit lanes, improving interchanges, or providing peak hour and plus lanes. Other suggestions for better use include making traffic flow smoother through surveillance or use of roundabouts, warning of dangerous locations and situations, in-car systems, and spreading traffic demand by informing road users of conditions.

2+1 Roads

The Netherlands also uses 2+1 roads as an interim condition until the road can be widened into a dual carriageway.[30] The 2+1 road is a single carriageway rural road with three lanes on which the center lane is used to provide overtaking opportunities in each direction. The guidelines published in 2008 were developed based on German and Swedish design standards. The implementation of the first 2+1 road (N50 between Kampen and the A28 motorway) was in 2007. Bearing the ultimate solution in mind, the designer is encouraged to design the 2+1 section as one of the carriageways and provide for a second carriageway in the overall design (especially ensuring that bridges, overpasses, etc., allow for this). This way, construction costs are spread while the design provides the necessary flexibility to introduce the second carriageway the moment traffic demand warrants it. Furthermore, the construction of the second carriageway causes minimal disruption to existing traffic.

Hard Shoulder

The first area to use hard shoulder lanes was the A27 roadway in Utrecht. When hard shoulders are used, they are monitored manually from a traffic control center and are operated through variable signs mounted on gantries. The following measures are in place when hard shoulder running is in operation:

- Maximum speed limit is reduced from 120 km/h (75 mi/h) to 90 km/h (56 mi/h).
- Emergency refuge areas are provided about every kilometer.
- Speed detection loops are placed every 500 to 600 m (1,640 to 1,970 ft) on normal lanes and every 75 m (246 ft) on the hard shoulder.
- Maximum spacing between gantries is about 700 m (2,300 ft).
- A green arrow is displayed when the lane is available for use and a red cross when it is closed.
- During hard shoulder running operation, the maximum speed limit is also displayed.
- Rescue vehicles are placed at junctions for rapid deployment in the event of an incident.[31]

Plus Lanes

The Netherlands also uses plus lanes in many areas, which are narrow lanes on the inside of the roadway that are opened at peak times to create capacity while at the same time speed limits are reduced to improve safety.[32,33] The plus lane is provided by reconstructing the existing roadway while keeping the hard shoulder. It is opened for travel use when traffic volumes reach levels that indicate congestion is growing.

Sweden

Higher Speed Roads

In Sweden, the objective is to decrease the number of people killed in crashes by 50 percent and severely injured by 25 percent from 2007 to 2020. The long-term "Zero vision" is that nobody complying with system laws (e.g., driving while sober and using a seatbelt, a modern car, and no excessive speeding) should be killed or severely injured.[34] According to their report, T. Bergh and M. Petersson[34] reported that to a great extent the design debate has focused on how to contribute to the safety objectives in a cost-effective way.

The following information on higher speed roadways was discussed:

- The first Swedish motorway (Malmo-Lund) was about 20 km (12.4 mi) and was opened in 1953 with a prewar German design without outer hard shoulders, with a narrow median, and without tapers at entries and exits. Most of the normally 110-km/h (68-mi/h) rural motorway system has a 26.5-m (87-ft) cross section (see figure 54). The fatality rate for 110-km/h (68-mi/h) motorways has decreased because of heavy implementation of median and side fences on older motorways.

  A political intervention based on environmental impact concern triggered a number of four-lane projects in the mid-1990s with an 18.5-m (61-ft) cross section (see figure 54). The crash rates for these projects. some 100 km (62 mi), "have not redeemed our expectations on safety performance." Figure 54 shows a compromise for new design with a 21.5-m (71-ft) cross section.

- The 2+1 cable barrier design has a cross section that is normally within a 13-m (43-ft) paved width (see figure 55). The first project opened in 1998 and was judged a major success by politicians and public opinion. More than 2,000 km are now open with the following distribution:
  - About 30 percent are 110-km/h (68-mi/h) semimotorways (i.e., grade separated, full access control).
  - About 30 percent are 90-km/h (56-mi/h) semimotorways.
  - About 30 percent are normal 90-km/h (56-mi/h) roads (i.e., with at-grade intersections, accesses, and vulnerable road users and slow-moving vehicles permitted).

Speed limits were changed to 100 km/h (62 mi/h) in 2008–2009 in the speed limit overview. The crash evaluation found an overall fatality rate almost equal to motorways, giving a reduction for some projects of 80 percent and over 50 percent including severe injuries, resulting in a savings of 50 lives a year. Main disadvantages are major barrier crash repairs.

- Major efforts are being concentrated on improved safety design on normal two-lane roads (6.6- to 11.5-m (21.6- to 37.7-ft) paved width) with a speed limit of 90 km/h (56 mi/h). These roads, some 10,000 km (6,213 mi), have about 90 fatalities a year, 40 percent in head- on and overtaking crashes and 25 percent in single roadway departure crashes. Average daily traffic is normally in the range of 1,000 to 5,000 vehicles. Three main alternatives are being examined:
  - Centerline rumble strips have been implemented on 4,000 km (2,485 mi) since 2005. The anticipated effect on fatalities and severe injuries is estimated at 10 to 15 percent based on Finnish and U.S. results, but those results have not been achieved yet in Sweden.

    The noise disturbance distance to avoid levels above 45 A-weighted decibels (dBA) has been extended to 150 m (492 ft) to occupied houses, giving an average of some 60 to 70 percent rumble strips on a project.
  - The rumble median with overtaking lanes concept includes the addition of overtaking sections at slightly longer intervals than in the 2+1 concept. A no-overtaking zone with a narrow rumble median is between the overtaking sections with information signs on the next overtaking opportunity.
  - Figure 56 illustrates the concept. Side safety fences are used in sharper outer bends. The costs of overlays, signs, and road markings are to be from maintenance money to facilitate a large-scale implementation over a reasonably short period.
  - A median barriers with overtaking lanes project was implemented in December 2009 at National Road 26 outside vaxjo (see figure 57 on next page). The 1+1 sections are squeezed into the existing 9-m (30-ft) width, with 0.75-m (2.5-ft) outer shoulders, 3.25-m (10.7-ft) traffic lanes, and a 1-m (3.3-ft) median.

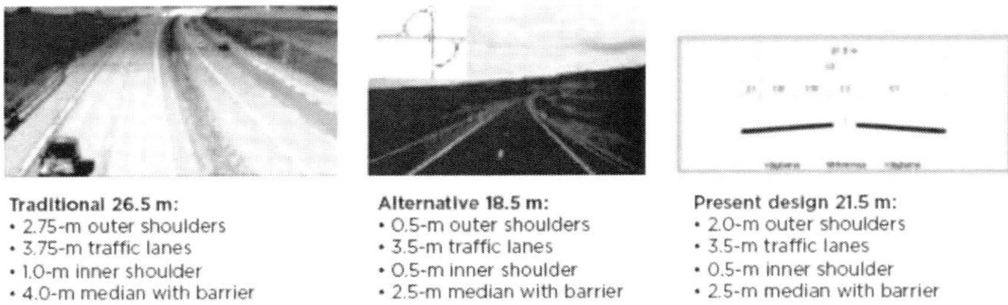

Figure 54. Sweden: motorway and alternative four-lane cross sections from the 1970s[34].

Figure 55. Sweden: typical 2+1 median barrier design on 13-m paved width[34].

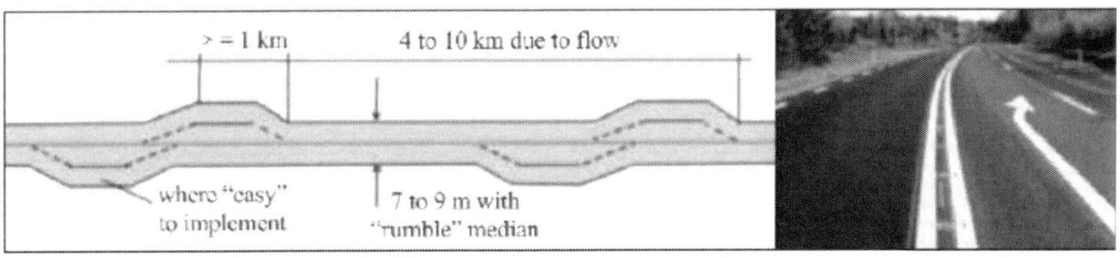

Figure 56. Sweden: overtaking lane principles with rumble or barrier median [34].

Figure 57. Sweden: median barrier project [34].

United Kingdom—England

Making Better Use

In his country report, J. Smart[35] discussed a strategy known as "Making Better Use." In 2000, after a government review of transport policy in the United Kingdom, the Highways Agency (highway authority for the strategic road network) was asked to develop a roads improvement strategy that would speed up delivery of large-scale road construction schemes over 10 years. The strategy became known as the 10-Year Plan. One element of the strategy became known as the Making Better Use program, which has a goal of creating additional capacity without moving outside the existing footprint of a highway. In addition, safety should not be compromised and environmental impact should be minimal. In geometric terms, this meant looking at the existing standard highway cross section and junction layouts to establish where additional lane widths could be developed or where using technology could assist in increasing capacity of existing lane configurations or enable lane control, lane entry, and lane designation changes to be made. This approach is to be considered for all levels of road hierarchy, whether the route is a single carriageway, dual carriageway, or motorway.

Wide Single 2+1

The wide single 2+1 is used when upgrading a single carriageway road into a dual carriageway is being considered. The single carriageway is marked in three lanes, with the center lane used for alternating directional overtaking sections (see figure 58). The existing 10-m (33-ft) width is used or combined with minor widening to 11 m (36 ft). Trials of this layout have been undertaken in the United Kingdom, and similar layouts operate in Finland, Germany, Ireland, and Sweden.[35]

Figure 58. United Kingdom: WS2+1 layout[35].

Active Traffic Management with Hard Shoulder Running

A trial project using the hard shoulders of the M42 roadway was conducted as a possible congestion relief solution. Motorists traveling southbound on the M42 at Junction 5 near Solihull were first to use the hard shoulder between motorway slip roads. The initiative created a continuous four-lane section of the motorway between Junctions 6 (Birmingham Airport) and 4 (Monkspath).[36] Initial results of the use of the shoulder on the M42 to relieve congestion showed that travel time was reduced by as much as 27 percent on weekdays.[37] Black and yellow electronic variable message signs inform motorists whether the hard shoulder is open to traffic and which lanes can be used to exit from or stay on the motorway.[36] The success of this trial resulted in the expansion of the program to the M6 roadway. Construction on the M6 to replace bridge joints, make surface repairs, and add lane markings are projected to be completed by spring 2011.[32]

Smart[35] provided additional details on the trial of the M42 section near Birmingham, England. The section is a three-lane motorway (six lanes plus full hard shoulders). Management techniques include mandatory variable speed limits, enhanced driver information signs, and a new congestion and incident management system. The system allows operators to open and close any lane on the motorway to traffic to help manage congestion at busy times of the day or during traffic buildup from an incident. Eventually, this will include using the hard shoulder as a running lane between junctions under controlled conditions. The infrastructure includes new lighting, gantries, electronic and static signing, emergency roadside telephones, emergency refuge areas, cameras, and mandatory variable speed limits. Use of the hard shoulder as an extra running lane during busy peak periods or incidents was introduced in September 2006 as the final phase in the active traffic management (ATM) project (see figure 59).[35] The key aspects of ATM are the following:

Figure 59. United Kingdom: hard shoulder use[38].

- Use of variable mandatory speed limits
- Dynamic use of the hard shoulder during periods of congestion or incidents
- Provision of dedicated emergency refuge areas for use when vehicles break down
- Installation of gantries with signals and variable message signs[38]

Managed Motorways

"Managed motorways" is a term used to encapsulate a range of techniques that can be employed to control the level and speed of traffic on a high-speed interurban route. According to Smart,[35] the managed motorway has developed in the United Kingdom from the successful trial of the ATM concept. The managed motorway concept is a context-sensitive solution being pursued to mitigate the detrimental effects of congestion while optimizing the investment that has or will be made in the roadway infrastructure. Agencies are using innovative geometric design treatments to optimize the use of the roadway cross section to actively manage traffic. They dynamically use operational strategies such as speed harmonization, vehicle restrictions (e.g., occupancy, trucks), or pricing and tolling to improve the performance of specific lanes or the entire motorway.

Truck Restrictions

The first truck restrictions in England were implemented on a pilot basis on northbound M42 near Warwickshire between 7 a.m. and 7 p.m. This 5-km (3-mi) stretch of roadway between Junction 10 (Tamworth) and Junction 11 (Appleby Magna) is only two lanes in each direction, and heavy vehicles, which make up 17 percent of the traffic, must travel slowly up the steep grade.[40] The ban prevents heavy trucks from overtaking each other and restricts them to one lane, leaving the other lane for faster moving vehicles. The intent is to enhance operations and safety and reduce congestion along this part of the roadway. Results of the ban, including safety impacts and travel time savings, are still unknown.[40]

Exclusive Bus Lanes on M4

The first bus lane on a motorway opened in 1997 on the M4 Spur motorway to Heathrow Airport. The Central Bus Station at the airport serves more than 1,600 buses per day, and the traffic during peak periods can be backed up the entire 1.4-km (0.9-mi) length of the spur.[41] The bus lane is identified by red surfacing treatment. Only buses, coaches, and emergency vehicles are allowed to use the bus lane; all other vehicles are prohibited. In 1999 a bus lane also opened on the M4 motorway on the eastbound lanes into London. The bus lane starts at Junction 3 and extends for 5.6 km (3.5 mi) to the point where the carriageway width reduces from three running lanes plus hard shoulder to two running lanes on the elevated motorway on Chiswick flyover. In this scheme, the offside running lane has been converted into a bus lane, reducing the number of lanes for other traffic to two lanes. This bus lane also has red surface treatment, but in this case taxis are also allowed to use it.[41]

United States

User Information Needs

In his presentation during the Workshop on Managed Motorways, M. Brewer discussed critical user information needs related to ATM.[42] Key questions include the following:

- Who is your driver audience?
    - Familiar (commuter) drivers versus unfamiliar (tourist) drivers
    - Long distance versus short distance
    - Commercial versus personal
- What do they need to know?
    - Motorway information (hours of operation, open lanes and shoulders, final destination, payment)
    - Traffic information (congestion, incidents, time savings)
    - vehicle information (occupancy, prohibitions, method of payment)
- When do they need to know it?

- Enough advance notice to process the information, determine if it applies, decide how to respond, and execute the response
- Earlier notice for unfamiliar or untrusting drivers
- How can you best convey that information?
  - No one-size-fits-all answer because ATM solutions can take many forms, but some common principles apply

Figure 60 shows a conceptualized model for driver decision. The model was developed as part of a Texas Department of Transportation project on managed lanes.[43] Figure 61 (see page 66) shows the relationship between the amount of information provided to drivers to process and the safety and efficiency of a driver's decisions. Drivers have limits to their ability to process information. Beyond that threshold, the safety and efficiency of drivers' decisions begin to degrade.

Identifying Locations with Sunglare

During the symposium, two papers were presented that discussed techniques for identifying locations with sunglare concerns.

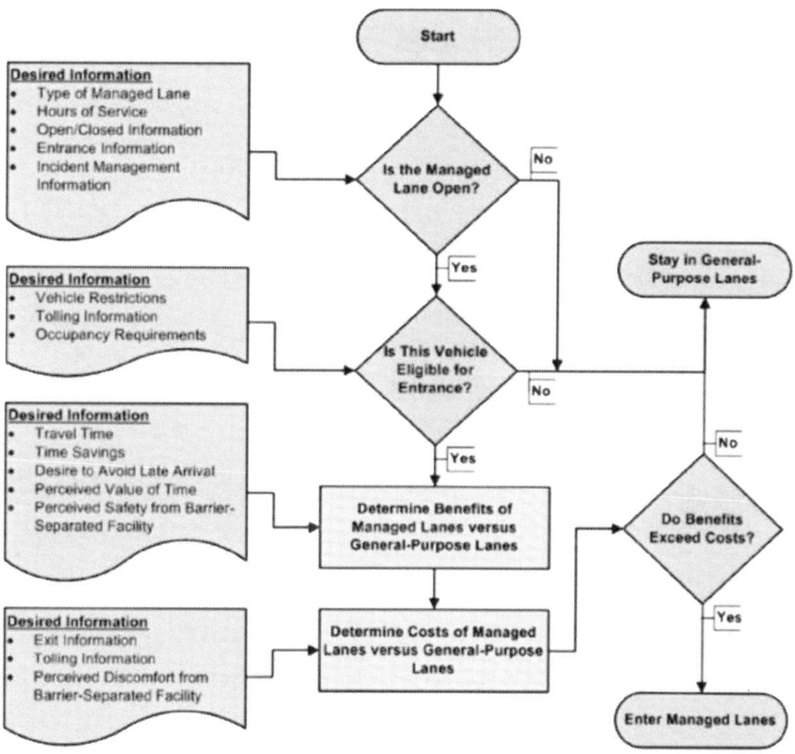

Figure 60. Conceptualized driver decision model[43].

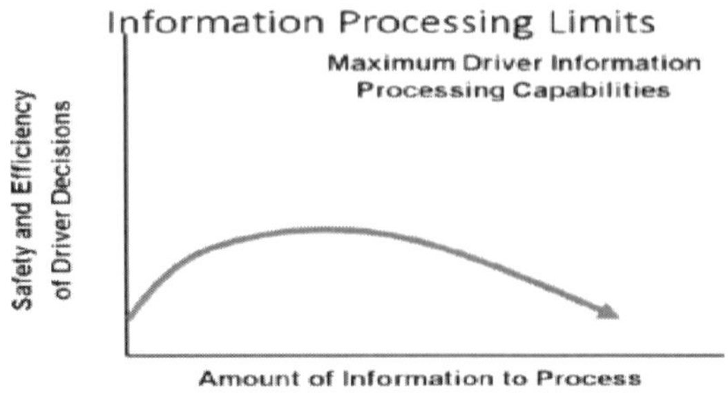

Figure 61. Relationship between efficiency and information to process[42].

Spain

The paper by R. Jurado-Pina et al[44] discusses a software tool developed at the Technical University of Madrid to identify and quantify driver vision impairment problems caused by sunglare and facilitate the design of countermeasures to prevent potential safety hazards. The computer program is based in a methodology developed at the university to determine the days and times of the year when sunglare may impair driver's vision on a particular road and the physical characteristics of its environment. The computer program has been implemented to represent the intervening variables in cylindrical charts and to perform the calculations to determine the times and days when drivers will experience sunglare vision impairment. The resulting software tool was applied in Spain in several studies of locations where sunglare could pose a safety hazard, including tunnel exits, freeway entrance ramps, and intersection approaches.

United States

A.M. Churchill and D.J. Lovell[45] of the University of Maryland discussed a procedure for auditing highway alignments for the effects of sunglare. They developed a mathematical method to determine, for a given geometrical description of a highway alignment, at what locations and times that alignment might be susceptible to sunglare problems. The calculations include astronomical algorithms to determine the vector direction of the sun from a given location on the highway at any time during the year, as well as the optical refinements necessary to account for atmospheric refraction. Essentially, the output of the model is a list of locations and times expected to suffer from sunglare effects. They noted that the algorithm could serve as one of a number of road safety audits conducted in a safety review for a given alignment, or it could serve as an input to an automated highway design process to account for safety deficiencies of candidate alignments.

Appendix C. Workshop on Managed Motorways— A Way Foward for the Future

Agenda

9:00 a.m. **Welcome—Opening Instructions**
John Smart, *Chartered Institution of Highways & Transportation, United Kingdom*

9:05 a.m. **From Freeways to Managed Motorways: A Generation of Change**
Jon Obenberger, *Federal Highway Administration, United States*

9:40 a.m. **Critical User Information Needs Related to Active Traffic Management**
Marcus Brewer, *Texas Transportation Institute, United States*

10:15 a.m. **Design Considerations and the Changes Imposed by Managed Motorway Operation**
Lucy Wickham, *Mouchel, and* **David Grant**, *Highways Agency, United Kingdom*

10:50 a.m. **Coffee Break**

11:05 a.m. **Planning and Implementing a Freeway Management System on Melbourne's M1**
John Cunningham, *VicRoads, Monash-CityLink Australia*

11:40 a.m. **Developments in Safety and Performance of Managed Motorways in the Netherlands**
Aad Wilmink, *Dutch Ministry of Transportation Netherlands*

12:15 p.m. **IT and Its Role as a Fundamental Element of Managed Motorways**
Matthew Clarke, *ATKINS, Highways and Transportation, United Kingdom*

12:50 p.m. **Panel Discussion**
MODERATOR: **John Smart**, *Chartered Institution of Highways & Transportation, United Kingdom*

All presenters and
Brian L. Ray, *Kittelson & Associates, Inc., United States*
Aniceto Zaragoza Ramirez, *Universidad Politecnica de Madridy Presidente de EUPAVE, Spain*
Charlie Howard, *Puget Sound Council of Governments, Seattle, Washington, United States*

1:30 p.m. **Adjourn**

APPENDIX D. HOST COUNTRY CONTACTS

Spain

Alfredo García, Ph.D.
Department of Transportation
Polytechnic University of valencia
Camino de vera, s/n
46071-valencia, Spain
Telephone: 011+34 963 871 374
Fax: 011+34 963 877 379
E-mail: argarcia@tra.upv.es

Ana Yago
Polytech University of valencia
Camino De vera, s/n
46071-valencia, Spain
Telephone: 011+34 963 871 374
Fax: 011+34 963 877 379
E-mail: ishagd4@hotmail.com

Jose A. Hinojosa
Telephone: 011+34 915 977 822
E-mail: jahinojosa@fomento.es

Vicente Ferrer Perez
Civil Engineer
Ministerio De Fomento
valencia, Spain
Telephone: 011+34 963 079 525
E-mail: vferrer@fomento.es

Jose Yuste Maizal
Civil Engineer
Ministry of Public Works
Valencia, Spain
Telephone: 011+34 963 079 538
Fax: 011+34 963 079 511
E-mail: jyuste@fomento.es

Minguel Mondria Garcia
Ingeniero do Caminos y Puertos
Director de Produccion y Calidad valencia, Spain
Telephone: 011+34 963 379 220 Fax: 011+34 963 371 429
E-mail: mmondria@typsa.es

Germany

Ulfert Joop
International Affairs Office
Federal Ministry of Transport,
Construction, and Urban Development
Robert-Schuman-Platz 1
D-53175 Bonn, Germany
Telephone: 011+49 228 300 5401
E-mail: ulfert.joop@bmvbs.bund.de

Konstantin Sauer
Referat StB 12
Federal Ministry of Transport, Construction, and Urban Development
Robert-Schuman-Platz 1
D-53175 Bonn, Germany
Telephone: 011+49 228 300 5120 or 5121
E-mail: ref-stb12@bmvbs.bund.de

Prof. Dr. Justin Geistefeldt
Institute for Traffic Engineering and Management
Ruhr-University Bochum
Universitaetsstr. 150
D-44801 Bochum, Germany
Telephone: +49 234 32-25936
Fax: +49 234 32-14151
E-mail: justin.geistefeldt@rub.de

Kerstin Lemke
Federal Highway Research Institute (BASt)
Bruderstrabe 53
D-51427 Bergisch Gladbach, Germany
Telephone: 011+49 2204 43-510

Fax: 011+49 2204 43-683
E-mail: lemke@bast.de

Reiner Dolger
Referat fur verkehrsanalyse und Forschung
Rheinland-Pfalz Ministerium fur Wirstschaft, verkehr,
Landwirtschaft und Weinbau Stiftsstrasse 9
D-55116 Mainz, Germany
Telephone: 011+49 06 130 162 283
Fax: 011+49 06 131 162 449
E-mail: reiner.doelger.mwvlw.rlp.de

Klaus Noll
Referent fur Strassenplanungen Rheinland-Pfalz
Ministerium fur Wirtschaft, verkehr
Landwirtschaft und Weinbau
Stiftsstrasse 9
55116 Mainz, Germany
Telephone: 011+49 06 131 162 209
Fax: 011+49 06 131 164 044
E-mail: klaus.noll@mwvlw.rlp.de

Helga Rottenau
Referat fur verkehrsanalyse und-forschung
verkehrsinformationssysteme, verkehrstelematik
Rheinland-Pfalz
Ministerium fur Wirtschaft
verkehr, Landwirtschaft Und Weinbau
Stiftsstrasse 9
55116 Mainz, Germany
Telephone: 011+49 06 131 162 171
Fax: 011+49 06 131 162 449
E-mail: helga.rottenau@mwv.rlp.de

Netherlands

Richard van der Elburg
Senior Advisor
Policy Analysis and Forecasts
Ministry of Transport, Public Works, and Water Management
Rijkswaterstaat Centre for Transport and Navigation
Schoemakerstraat 97, PO Box 5044
2600 GA Delft, Netherlands
Telephone: 011+31 887 982 372
Fax: 011+31 887 982 999
E-mail: richard.vander.elburg@rws.nl

Bert Hellerman
Senior Advisor
Rijkswaterstaat
Ministry of Transport, Public Works, and Water Management
Rijkswaterstaat Center for Transport and Navigation
Schoemaderstraat 97c
2628 vk Delft
PO Box 5044
2600 GA Delft, Netherlands
Telephone: 011+31 08 87 98 24 78
Fax: 011+31 08 87 98 29 99
E-mail: bert.helleman@rws.nl

Alex van Loon
Advisor, Policy Analysis and Forecasts
Ministry of Transport, Public Works, and Water Management
Rijkswaterstaat Centre for Transport and Navigation
Schoemakerstraat 97, PO Box 5044
2600 GA Delft, Netherlands
Telephone: 011+31 08 87 98 22 90

Fax: 011+31 08 87 98 29 99
E-mail: alex.van.loon@rws.nl

Marthe van Dongen, Ph.D.
Senior adviseur Trace
m.e.r.-centrum
Afdeling Inpassing Netwerken
Rijkswaterstaat
Dienst verkeer en Scheepvaart
van der Burghweg 1
Postbus 5044
2600 GA Delft, Netherlands
Telephone: 011+31 08 87 98 23 45
Fax: 011+31 08 87 89 29 99
E-mail: marthe.van.dongen@rws.nl

Gerald Uittenbogerd
DvS Consultant
E-mail: gerald.uittenbogerd@rws.nl

Huib Kwint
Advisor
Rijkswaterstaat
Ministry of Transport, Public Works, and Water Management
Rijkswaterstaat Center for Transport and Navigation
Van der Burghweg 1
2628 CS Delft, Netherlands
PO Box 5044
2600 GA Delft, Netherlands
Telephone: 011+31 08 87 98 22 89
Fax: 011+31 08 87 98 29 99
E-mail: huib.kwint@rws.nl

Roel Nijsten
Team Leader Transportation/EIA centre
Section Landscaping and Environment
Rijkswaterstaat Centre for Transport and Navigation
van der Burghweg 1
PO Box 5044
2600 GA Delft, Netherlands
Telephone: 011+31 08 87 98 23 57
Fax: 011+31 08 87 98 29 99
E-mail: roel.nijsten@rws.nl

United Kingdom

Paul Johnson
Head of Research Strategy
Network Services Directorate
Highways Agency
The Cube, 199 Wharfside St.
Birmingham B1 1RN, United Kingdom
Telephone: 011+44 121 678 8275
E-mail: paul.johnson@highways. gsi.gov.uk

Clair Griffin
Highways Agency
The Cube, 199 Wharfside St.
Birmingham B1 1RN, United Kingdom
Telephone: 011+44 121 678 8480 Fax: 011+44 121 678 8108
E-mail: clare.griffin@highways.gsi. gov.uk

David Grant
Group Manager
Highways Agency, The Cube

199 Wharfside St.
Birmingham B1 1RN, United Kingdom
Telephone: 011+44 07 802 919 582
E-mail: david.grant@highways.gsi. gov.uk

Martin Stott
Operations Manager
Highways Agency
West Midlands Regional Control Center
Quinton Business Park, 1 Ridgeway
Birmingham B32 1AF, United Kingdom
Telephone: 011+44 121 421 8851
Fax: 011+44 121 421 8849
E-mail: martin.stott@highways. gsi.gov.uk

Mike Wilson
Regional Director–East Midlands
Telephone: 011+44 121 678 8055
E-mail: mike.wilson@highways.gsi.gov.uk

Nick Hopcraft
Divisional Director, Managed Motorways
Highways Agency
C3 5 Broadway

Birmingham, B15 1 BL, United Kingdom
Telephone: 011+44 121 678 8342
Fax: 011+44 121 678 8211
E-mail: nick.hopcraft@highways. gsi.gov.uk

Max Brown
Senior Traffic Technology Officer Network Services Directorate,
Research and Operational Guidance Division
Highways Agency
Temple Quay House
2 Square Temple Quay
Bristol BS1 6HA, United Kingdom
Telephone: 011+44 117 372 8541
Fax: 011+44 117 372 8810
E-mail: max.brown@highways.gsi. gov.uk

Vicky Bowden
Assistant Knowledge Strategy Coordinator
Highways Agency
403 City Tower, Piccadilly Plaza
Manchester, M1 4BE, United Kingdom
Telephone: 011+44 161 930 5809
Fax: 011+44 161 930 5256
E-mail: vicky.bowden@highways. gsi.gov.uk

Ruth Tilstone
Highways Agency
403 City Tower, Piccadilly Plaza
Manchester, M1 4BE, United Kingdom
Telephone: 011+44 161 930 5809
Fax: 011+161 930 5256
E-mail: ruth.tilstone@highways.gsi.gov.uk

Paul Unwin
Senior Project Manager
Highways Agency
The Cube, 199 Wharfside St.
Birmingham B1 1RN, United Kingdom
Telephone: 011+44 121 678 8180
E-mail: paul.unwin@highways.gsi.gov.uk

Brian Barton, Bsc. CEng.MICE
NetServ, Technical Services Division
Highways Agency
3C Federated House
London Road, Dorking

Surrey RH4 1SZ, United Kingdom
Telephone: 011+44 130 687 8292
Fax: 011+44 130 687 8301
E-mail: brian.barton@highways. gsi.gov.uk

Sarah Garland
Managed Motorway
Delivery Office (Operations and HOv Lanes)
Telephone: 011+44 121 087 4161
E-mail: sarah.garland@highways. gsi.gov.uk

Andrew Page-Dove
Program Manager–Managed
Motorways (Operations)
Telephone: 011+44 781 807 7913
E-mail: andrew.page-dove@highways.gsi.gov.uk

Lucy Wickham, CEng MICE MIHT
Technical Director–Transport Operations & Technology
Mouchel
Export House, Cawsey Way Woking
Surrey GU21 6QX, United Kingdom
Telephone: 011+44 148 373 1122
E-mail: www.mouchel.com

Appendix E. Amplifying Questions

Scan Purpose

Continued growth in travel along congested urban freeway corridors exceeds the ability of agencies to provide sufficient roadway capacity in major metropolitan areas based on limited public funding for roadway expansion and improvement projects. High construction costs, constrained right-of-way, and environmental factors are pushing agencies to explore solutions, such as active traffic management or managed lanes, to maximize throughput under congested conditions and improve safety to reduce collisions and nonrecurring congestion.

The purpose of this scan is to examine innovative geometric design practices and techniques used in other countries to improve the operational performance of congested freeway facilities without compromising safety. Finding cost-effective options to mitigate traffic congestion on urban freeway facilities is one of the most significant challenges State departments of transportation and regional transportation organizations face. Internationally, transport agencies are using geometric design treatments linked to operational strategies, such as reassigning the roadway cross section, to dynamically reduce congestion while maintaining or improving the safety performance of freeways in congested urban freeway corridors.

Amplifying Questions

A. Geometric Design Practices

The questions in this section are targeted to geometric design practices used to optimize the performance of existing or future freeway capacity. Examples of geometric design practices include special-use lanes, such as high-occupancy vehicle (HOV) or exclusive truck lanes; value-priced or high-occupancy toll (HOT) lanes; and reserved areas for vehicle refuge, enforcement, or incident response and recovery. Other examples of geometric design practices include temporary use of shoulders, lane restrictions, and access control. Also to be reviewed during the scanning study are signing, pavement marking, traffic control, lighting, speed, and other elements considered in the design.

1. In the United States, the prevailing document used for highway geometrics is the American Association of State Highway and Transportation Officials (AASHTO) *A Policy on Geometric Design of Highways and Streets,* also known as the *Green Book.*
   a. What document is established that sets your national design criteria?
   b. Do you recognize AASHTO and the guidance given in the Green Book and, if so, in what capacity?
2. What practices have been used in your country to support active traffic management? Were they successful or unsuccessful? Why?
3. What cross sections and dimensions are used for the different geometric design practices? Can you provide example schematics?
   a. For example, what are the acceptable shoulder widths? Does it vary based on the length of the shoulder encroachment? In other words, can you get by with a narrower shoulder for a short distance (such as a bridge), but need wider shoulders for longer roadway segments? Are there any increments of shoulder width that are avoided (such as avoiding a 6 foot (1.83 meter) inside shoulder because it is too narrow to safely accommodate a disabled motorist, but not narrow enough to discourage motorists to move to the outside shoulder for refuge)?
   b. For example, when considering lane widthreductions below standard values, what geometric factors and other elements are reviewed to make the decision, (e.g., tangent-versus-curvy alignment, percent trucks)?
4. Where constraints require the use of dimensions for cross section elements that are less than desired, is there an order in which the elements are prioritized for reduction? For example, one might reduce a buffer width between mixed and managed lanes first, reduce left shoulder width second, reduce lane width third, etc.
5. What types of geometric modifications to other-than-cross-section features of existing freeways have proven necessary where active traffic management is deployed? For example, how is ramp design affected (i.e., acceleration or deceleration lengths, location of junctions, where and how an HOv lane is accessed, changes needed because ramp is located along side of a dynamic shoulder)? Another example could be overhead or roadside sign support locations. Specific geometric details and/or specific roadway examples would be of interest.

6. Is it inherent with the use of alternative geometric design practices—particularly dynamic shoulder use and managed lanes— that features such as lighting, signing, and delineation need to be enhanced beyond typical provision? If so, to what extent?
7. Is enforcement provided in a separate area for monitoring and citation issuance?
8. What are practitioners' thoughts and practices on the use of less-than-desired dimensions, such as the use of narrower-than-standard lane widths? These practices are likely to be proposed as retrofits to existing corridors with space restrictions. The navigational and decision demands associated with them would most likely compound demands related to substandard geometric elements.
9. Under what circumstances has or would dynamic shoulder use be considered? Adding lane capacity to main lanes is the obvious application, but has or can the strategy be used to optimize operations in other ways (e.g., a one-lane versus two-lane ramp exit for offpeak versus peak-hour demand conditions)?
   a. Are weather issues such as rainwater runoff or spread and snowmelt considered before running traffic on shoulders?
   b. Is traffic running on the shoulders allowed during daylight hours only? If it is allowed at night, are these sections lighted?
10. Has deployment of the alternative geometric design practices caused designers and jurisdictions to reduce design speeds in these corridors?
11. For restriping retrofit applications, is cross slope broken on lane lines or in between?
12. Have there been any studies or simulations to evaluate driver reading comprehension of complex signing? Do you have examples of signing (both static and dynamic) used with active traffic management?
13. Please address the following on vehicle type:
    a. What kind of vehicle-type restrictions, if any, have been instituted on corridors where these treatments have been employed?
    b. Have any geometric design strategies (e.g., narrow lanes) been particularly problematic for trucks?
    c. Are there any innovative design practices to accommodate operation of buses?
14. Are there any innovative design practices to accommodate congestion pricing facilities?

B. Performance Measures

The following questions focus on performance measures (and/or metrics) to evaluate, monitor, and report on the performance of geometric design alternatives aimed at improving the performance and flexible use of freeway facilities (e.g., full-time capacity addition, part-time capacity addition). The scan team seeks information on existing and proposed methods, procedures, tools, public outreach, and techniques used to assess safety and operational implications or compare geometric design alternatives, active traffic management strategies, and innovative uses of the cross section of a freeway.

1. Before implementing new geometric design practices, did your agency undertake any formal or informal evaluation of the potential risks in the design process? If so, please explain the evaluation process used.
2. Have any studies been performed to evaluate the operational and safety changes experienced after implementing active traffic management and/or specific geometric

design alternatives (e.g., cross section tradeoffs that support active traffic management)? If so, please describe.

   a. For example, if you have tried strategies that involve using the shoulder as an additional travel lane during certain peak times, have you performed any safety studies to assess the effects of such strategies?

3. What relationships have you found between geometric design practices and active traffic management or operational strategies?
4. Has your agency developed any specific performance goals for evaluating active traffic management and/or specific geometric design practices?
5. What performance measures do you use to evaluate the effects of geometric design alternatives used in active traffic management on your roadways?
   a. Do you use some form of level of service, such as number of congested hours, travel speed, or amount of delay?
   b. Do these measures show clear improvements in safety, operations, and maintenance?
   c. What types of alternatives appear to provide the best benefits?
   d. Do you apply a lessons-learned approach to help refine existing alternatives and develop new ones?
6. How do you face the challenge of balancing your project between designing for safety and designing for increased mobility?
   a. Do the decisions you make have differing effects on safety and mobility?
   b. How do you quantify the effects on safety and mobility?
   c. What risk management tools and processes have been used to understand these tradeoffs in planning and design and make decisions?
7. Have you been successful in implementing innovative technologies into your overall traffic operations program? Specifically, have you developed control devices other than the traditional items, such as signs, markings, and signals?
8. What role do traffic signal control, signage, metering of roadways, and alternate route selection play in managing your overall capacity?
9. How do you relate traveler information or mobility management to the placement of traffic restrictions or pricing in dealing with operational strategies?
10. How are speed limits determined?
    a. How are speed limits determined for sections with geometric designs that have been modified to accommodate active traffic management?
    b. Are speed limits considered when selecting or developing geometric design alternatives?
    c. Is compliance with speed limits good or poor?
    d. Where local laws do not allow dynamic signing to dictate regulated speed posting, have advisory speed messages been effective in controlling or influencing travel speeds?

C. Planning

The following questions focus on the planning components used to integrate geometric design practices for active traffic management strategies. The scan team is interested in institutional issues experienced, such as political involvement; public acceptance or controversy; and organizational impacts, such as the need for addressing competencies, processes, structures, or other resource and leadership matters.

1. What are the typical steps in initiating your project development process?
2. Please describe the procedures used for defining and developing projects from a conceptual through final design stage?
3. How are improvements to motorway lanes planned? Who undertakes that planning? At what points in your plan development process do you begin to look at various forms of active management strategies? When do you consider the geometric design alternatives that need to be in place to use the active management strategies?
4. Have you made changes in your planning or design process to better accommodate active traffic management?
5. Please describe how policy decisions are made when using geometric design criteria to fit the overall context of a project.
6. How do you address public involvement in your project development?
   a. Specifically, how do you articulate the benefits of alternative geometric design techniques?
   b. What role do both the public and media play in political perceptions?
   c. Are particular issues raised by decisionmakers and, if so, how have they been addressed?
7. Have these corridorwide applications been deployed as a limited-length interim approach to getting more life out of existing roads in advance of a larger fix, or are they used as long-term, permanent corridor reinventions— or both?
8. Is there an institutional process to ensure the consideration or inclusion of active traffic management in developing long-range regional plans?
9. What techniques have been used to integrate enforcement and traffic incident management needs into planning and design?
10. What specialized training programs have you developed to accommodate your innovative projects or technologies?
11. Active traffic management requires a long-term commitment to operations, as well as to maintenance, enforcement and patrolling, and traffic management. How has this commitment been planned and budgeted for?

D. Benefits and Lessons Learned

The following questions focus on the benefits of and lessons learned from different types of geometric design practices used to optimize the performance and flexible use of existing or future expanded freeway capacity.

1. What types of mitigation measures to minimize safety risks were implemented in conjunction with geometric design practices implemented as part of active traffic management strategies?
2. What tools and methods of communicating the benefits have been developed and how have they been implemented?
3. In planning or after the fact, have the owning agencies explored the life-cycle costs of these alternatives (including management, maintenance, increased enforcement),

and have they computed life-cycle benefit/cost ratios for comparison with more conventional freeway strategies?

4. Should agencies consider these types of innovative strategies for deployment only selectively on their cities' freeway systems, or are they considered viable for application systemwide?
5. For the concept of self-explaining roads, what level of study and research has been completed in your country to demonstrate which geometric design elements and traffic control devices are most effective at conveying the proper message to drivers?

GLOSSARY

Active traffic management. Active traffic management is an approach to dynamically manage and control traffic demand and available capacity of transportation facilities based on prevailing traffic conditions, using one or a combination of real-time and predictive operational strategies. When implemented with traditional traffic demand management strategies, these operational strategies help maximize the effectiveness and efficiency of the transportation facility and result in improved safety, trip reliability, and throughput. A truly active management philosophy dictates that the full range of available operational strategies be considered, including the various ways these strategies can be integrated with existing infrastructure, to actively manage the transportation system to achieve system performance goals. This includes traditional traffic management and intelligent transportation system (ITS) technologies, as well as new technologies and nontraditional traffic management technologies used in other parts of the world.

Carriageway. The part of a main road used for vehicles, especially one side of a major two-way highway carrying traffic in one direction only.

Dynamic rerouting. The provision of route information on overhead sign gantries along a roadway in response to recurrent and nonrecurrent congestion. The signs provide en route guidance information to motorists on queues, major incidents, and appropriate routes.

Dynamic message sign. A permanently installed or portable electronic traffic sign used on roadways to give travelers information about roadway conditions, including traffic congestion, crashes, incidents, work zones, speed limits, alternate routes, or special events on a specific highway segment. It can be changed or switched on or off as required and can be used to provide roadway lane control, speed control, and operational restrictions. Also known as a **changeable message sign** or **variable message sign.**

Hard shoulder running. See temporary shoulder use.

Line control or lane control. Line or lane control is the procedure of controlling the use of a lane through signs mounted on a gantry. The variable lane control sign shows a green arrow when the lane is open, a yellow arrow when the lane is closing, and a red X when the lane is closed to traffic.

Managed lanes. Highway facilities or a set of lanes in which operational strategies are implemented and managed (in real time) in response to changing conditions to preserve unimpeded flow. They are distinguished from traditional lane management strategies in that they are proactively implemented and managed and may involve using more than one operational strategy with the goal of achieving unimpeded flow.

Managed motorways. The managed motorway concept includes a combination of active or dynamically managed operational regimes, specific designs of infrastructure, and technology solutions. It uses a range of traffic management measures to actively monitor the motorway and, based on the monitoring, dynamically control speeds, add capacity, and inform road users of conditions on the network with the objective of optimizing traffic and safety performance.

Merge control. A variation of the temporary shoulder used in Germany. Typically, it is applied at entrance ramps or merge points where the number of downstream lanes is fewer than the number of upstream lanes. Lane control signals are installed over both upstream approaches before a merge. They provide priority to the facility with the higher volume and give a lane drop to the lesser volume roadway or approach. Also known as **junction control** or **mainline merging control.**

Motorway. A limited-access road intended for traveling relatively fast over long distances. A term commonly used in Europe, a motorway in the United States is known as a freeway.

Plus lane. The practice of opening up the shoulder next to the inside lane of traffic for temporary use to address capacity bottlenecks on the freeway network during times of congestion and reduced travel speeds. Travel on the shoulder is permitted only when speed harmonization is active and speed limits are reduced. Signs indicate when travel on the shoulder is permitted.

Queue warning. The display of warning signs and flashing lights along a roadway to alert that congestion and queues are ahead.

Ramp metering. Procedures used to reduce congestion by managing vehicle flow from local- access on-ramps. The entrance ramp is equipped with a traffic signal that allows vehicles to enter the freeway at predetermined intervals.

Smarter highways. Term used by the Washington State Department of Transportation to describe implementation of active traffic management to increase roadway efficiency and help drivers travel in a safer and smarter manner. Smarter highways include overhead signs that display variable speed limits, lane status, and real-time traffic information.

Speed harmonization. The practice of using an expert system to monitor data coming from field- deployed sensors on a roadway and automatically adjust speed limits when congestion thresholds are exceeded and congestion and queue formation are impending. Sign gantries that span the facility provide speed limits and additional information, depending on roadway conditions.

Symbology. The use of graphic symbols to represent information pertinent to roadway users. The European practice of using symbology follows the Vienna Convention.

Temporary shoulder use. The practice of opening up the shoulder next to the outside lane of traffic for temporary use to address capacity bottlenecks on the freeway network during times of congestion and reduced travel speeds. Travel on the shoulder is permitted only when speed harmonization is active and speed limits are reduced. Signs indicate when travel on the shoulder is permitted. Also known as **hard shoulder running** or a **rush-hour lane.**

Truck restrictions. Any restrictions along a roadway on the operation of trucks or heavy goods vehicles. Examples include restricting trucks to specific lanes, prohibiting them from using particular lanes, limiting their operating speed, or prohibiting their use of the entire facility during specific periods of the day.

Variable speed limits. Speed limits that change based on road, traffic, or weather conditions. Also known as **dynamic speed limits.**

REFERENCES

[1] "International Programs" (2008). Federal Highway Administration Web site. Available at http://international.fhwa.dot.gov/tmpl.cfm?title= scanning.

[2] "International Technology Scanning Programs" (2010). Federal Highway Administration Web site. Available at http://international. fhwa.dot.gov/scan/.

[3] Clark, E. I. M. (2010). "*IT and Its Role as a Fundamental Element of Managed Motorways*". Transportation Research Board 4th International Symposium on Highway Geometric Design, Managed Motorway Workshop. Valencia, Spain.

[4] Hinojosa, J. A., Perez, v. F. & Maical, V. (2010). "*Gestion De Autopistas.*" Presentation to Freeway Geometric Design Scan Team, June 4.

[5] Google Earth. Map Data.

[6] Lemke, K. (2010). "*Hard Shoulder Running as a Short-Term Measure to Reduce Congestion*" Transportation Research Board 4th International Symposium on Highway Geometric Design. Valencia, Spain.

[7] Lemke, K. (2010). "*Hard Shoulder Running as a Short-Term Measure to Reduce Congestion*". Transportation Research Board 4th International Symposium on Highway Geometric Design. Valencia, Spain.

[8] Geistefeldt, J. (2010). "*Traffic Management and ITS in Hessen.*" Hessian Road and Traffic Authority. Presentation to Freeway Geometric Design Scan Team, June. Frankfurt, Germany.

[9] Hessian State Office for Road and Traffic Affairs. (2009). "*Congestion-Free Hessen 2015: A Success Story*". Available at www. staufreieshessen2015.de/irj/servlet/prt/

portal/prtroot/slimp.CMReader/HMWvL_15/Staufrei_Internet/med/9da/9da0999c-199c-021f-3efe-f 91921321b2c,22222222-2222-2222-2222-222222222222.

[10] German "*Duration of Congestion on Motorways in Hessen*" pamphlet (undated). Project Management, Congestion-Free Hessen 2015.

[11] Hessisches Landesamt fur StraBen-und Verkehrswesen (2010). "*Temporary use of hard shoulders*". Staufreies Hessen 2015. Available at www.staufreieshessen2015.de/irj/Staufrei_Internet?rid=HMWvL_15/Staufrei_Internet/nav/09f/09f20b18-af71-ff11-f3efef91921321b2,f4c7b18a-f71f-f11f-3efef91921321b2c,,,11111111-2222-3333-4444-100000005003percent26overview=true.htm&uid=09f20b18-af71-ff11-f3ef-ef91921321b2.

[12] Dolger, R. (2010). "*FGD Scanning Tour, Welcome to Rheinland-Pfalz!*" Referat fur verkehrsanalyse und Forschung, Rheinland-Pfalz Ministerium fur Wirstschaft. Presentation to Freeway Geometric Design Scan Team, June. Mainz, Germany.

[13] Federal Ministry of Transport. (2000). Unfallgeschehen im Bereich von Streckenbeeinfl ussungsanlagen unter besonderer Berücksichtigung der verkehrsbelastung.

[14] RAA. Richtlinien fuer die Anlage von Autobahnen. (2008).

[15] Hellernan, B. (2010). "*Managed Motorways in the Netherlands.*" Centre for Transport and Navigation. Presentation to Freeway Geometric Design Scan Team, June 8.

[16] Rijkswaterstaat. "*Dynamic Speed Limits*" pamphlet. (undated).

[17] Strookbreedtes voor gestrekte wegvakken (hoofdrijbanen en parallelrijbanen). (2007).

[18] Johnson, P. (2010). "*Welcome and Introduction.*" Presentation to Freeway Geometric Design Scan Team, June 10.

[19] Brown, M. (2010). "*Managed Motorways: Design Standards and Technology.*" Presentation to Freeway Geometric Design Scan Team, June 10.

[20] *Advanced Motorway Signaling and Traffic Management Feasibility Study: A Report to the Secretary of State for Transport.* (March 2008). Available at www.dft.gov.uk/pgr/roads/network/policy/mtorsigntraf management/advancedmotorwaysignal.pdf.

[21] Hopcraft, N. (2010). "*Managed Motorways Overview and Introduction.*" Presentation to Freeway Geometric Design Scan Team, June 10.

[22] Barton, B. (2010). "*Procedures, Tools, Public Outreach Tools to Assess Safety or Compare Design Alternatives.*" Presentation to Freeway Geometric Design Scan Team, June 10.

[23] Wickham, L. (2010). "*Questions and Answers Session.*" Discussion with Freeway Geometric Design Scan Team, June 10.

[24] Unwin, P. (2010). "*Managed Motorways.*" Presentation to Freeway Geometric Design Scan Team, June 11.

[25] Stott, M. (2010). "*Traffic Officers.*" Presentation to Freeway Geometric Design Scan Team, June 11.

[26] Cunningham, J. (2010). "*Planning and Implementing a Freeway Management System on Melbourne's M1.*" Presentation to Freeway Geometric Design Scan Team During Workshop on Managed Motorways—A Way Forward for the Future, June 2. Valencia, Spain.

[27] Arndt, O. K., Cox, R. L., Lennie, S. C. & Whitehead, M. T. (2010). "*Provision of Sight Distance Around Concrete Barriers and Structures on Freeways and Interchanges*". Transportation Research Board 4th International Symposium on Highway Geometric Design. valencia, Spain.

[28] Kjemtrup, K. (2010). "*Country Report—Denmark*". Transportation Research Board 4th International Symposium on Highway Geometric Design. Valencia, Spain.

[29] Greibe, P. (2010). "*Chevron Markings on Freeways: Effect on Speed, Gap, and Safety*". Transportation Research Board 4th International Symposium on Highway Geometric Design. Valencia, Spain.

[30] Schermers, G., Wegman, F. van vliet, P. van der Horst, R. & Boender, J. (2010). "*Country Report— the Netherlands*". Transportation Research Board 4th International Symposium on Highway Geometric Design. valencia, Spain.

[31] Department for Transport. *Annex to the Department for Transport Advanced Motorway Signaling and Traffic Management Feasibility Study Report.* (undated). Accessed at Department for Transport Web site: www.dft.gov.uk/pgr/.

[32] Whitelaw, J. (2009). "Managed Motorways: Refining the Art" (September 24). *The New Civil Engineer.* Accessed at New Civil Engineer Web site: www.nce.co.uk.

[33] Mirshahi, M., Obenberer, J. Fuhs, C. Howard, C. E. Krammes, R. A. Kuhn, B. T. Mayhew, R. M. Moore, M. A. Sahebjam, K. Stone, C. J. & Yung, J. L. (2007). *Active Traffic Management: The Next Step in Congestion Management.* FHWAPL-07-012, Federal Highway Administration, U.S. Department of Transportation, Washington, DC.

[34] Bergh, T., & Petersson, M. (2010). "*Country Report— Sweden*". Transportation Research Board 4th International Symposium on Highway Geometric Design. Valencia, Spain.

[35] Smart, J. (2010). "Country Report—United Kingdom". Transportation Research Board 4th International Symposium on Highway Geometric Design. Valencia, Spain.

[36] Transport News. (2008) "*Drivers to use M6 hard shoulder to ease congestion*" (Sept. 10). Birmingham Post. Accessed at www.birmingh ampost.net.

[37] "Public Exhibitions Announced for M1 J10 to J13 Improvement Scheme" (March 19, 2009). *COI News Distribution Service Release.* Accessed at www.highways.gov.uk/ news/pressrelease.aspx?press releaseid =171923.

[38] MacDonald, M. (2008) "*ATM Monitoring and Evaluation, 4-Lane variable Mandatory Speed Limits, 12-Month Report (Primary and Secondary Indicator)*" (June 30). Highways Agency. Accessed at www.dft.gov.uk/pgr/roads/tpm/ m42activetraffic management/atm12 mthsumrep.pdf.

[39] "What Happens If I Break Down." Highways Agency Web site: www. highways.gov. uk/roads/ projects/4690.aspx.

[40] "*Lorries banned from overtaking to cut congestion on three miles of the northbound M42 in Warwiskchire*" (undated). Highways Agency Web site: www.highways.gov.uk/ news/press_releases/general/2005_10_03d. htm.

[41] White, C., & MacDonald, M. (1999). "*Motorway Bus Lanes—A False Dawn or the Start of A Revolution*". UK 249, poster session of the European Transport Conference.

[42] Brewer, M. (2010). "*Critical User Information Needs Related to Active Traffic Management.*" Presentation to Freeway Geometric Design Scan Team, June 1.

[43] Kuhn, B., Goodin, G. D., Ballard, A., Brewer, M., Brydia, R., Carson, J., Chrysler, S., Collier, T., Eisele, W., Fenno, D., Fitzpatrick, K., Ullman, B., Ullman, G., & venglar, S. (2005). *Findings from Texas: Five Years of Research on Managed Lanes.* FHWA/ TX-06/0-4160-25.

[44] Jurado-Pina, R., Pardillo-Mayora, J. M. & Puy-Huarte, J. (2010). "*Software Tool for the Analysis of Highway Alignments to Detect and Prevent Sun Glare vision Impairment*

Hazards". Transportation Research Board 4th International Symposium on Highway Geometric Design. Valencia, Spain.

[45] Churchill, A. M., & Lovell, D. J. (2010). "*A Procedure for Auditing Highway Alignments for the Effects of Sun Glare*". Transportation Research Board 4th International Symposium on Highway Geometric Design. Valencia, Spain.

In: Managed Motorways
Editors: Elvis L. Duncan and Michael J. Snyder
ISBN: 978-1-61942-881-2

Chapter 2

REDUCING CONGESTION AND FUNDING TRANSPORTATION USING ROAD PRICING IN EUROPE AND SINGAPORE*

United States Department of Transportation

ACRONYMS AND ABBREVIATIONS

AASHTO	American Association of State Highway and Transportation Officials
ANPR	automated number plate recognition
BAG	German toll enforcement vehicles
CBD	central business district
CO_2	carbon dioxide
COE	Certificate of Entitlement
CRD	Congestion Reduction Demonstration
CURACAO	Coordination of Urban Road-User Charging Organizational Issues
CZK	Czech koruna
DOT	department of transportation
DSRC	dedicated short-range communications
EETS	European Electronic Toll Service
ERP	Electronic Road Pricing
EU	European Union
EUR	euro
FHWA	Federal Highway Administration
GBP	U.K. pound
GmbH	German legal designation for company incorporation
GPS	Global Positioning System
GSM	Global System for Mobile Communications

* This is an edited, reformatted and augmented version of the U.S. Department of Transportation, Federal Highway Administration report, dated December 2010.

HGV	heavy goods vehicle
HOT lane	high-occupancy toll lane
ITS	intelligent transportation systems
IU	in-vehicle unit
km/h	kilometers per hour
LTA	Land Transport Authority
OBU	onboard unit
PCU	passenger car unit
PPP	public-private partnership
SEK	Swedish krona
SGD	Singapore dollar
TfL	Transport for London
USD	U.S. dollar

EXECUTIVE SUMMARY

The evidence in the countries visited in this scan shows that road pricing can play a vital role in creating new funding for transportation, encouraging improved quality of life in the urban environment, advancing economic productivity for goods movement and business, increasing the use of public transit, and reducing congestion and emissions.

Road Pricing

Road pricing has a long history in the form of tolled bridges, tunnels, and turnpikes designed to generate revenue to pay for the construction, operation, and maintenance of these facilities. In the last half century, road pricing has been viewed as an opportunity to leverage the principles of supply and demand to manage traffic with congestion pricing. This is achieved by charging drivers a user fee (i.e., a toll or other charge) that may vary by traffic demand, time of day, vehicle classification, or other factors. In practice, road pricing provides a tool for road operators to manage limited roadway capacity to reduce congestion and maintain free-flow traffic conditions on highways, as well as to generate revenues that help pay capital, operating, and maintenance costs.

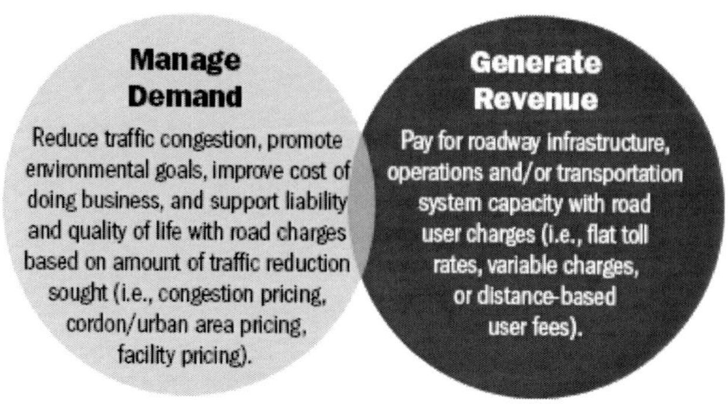

Figure 1. Purposes of road pricing.

The broad application of congestion pricing in the United States has been limited because of political, institutional, and public acceptance concerns. However, variable charges have been used successfully by many U.S. industries, including hospitality, air travel, utilities, and telecommunications.

Road pricing has been instituted on a broader basis in other countries, notably the Czech Republic, Germany, Singapore, Sweden, and the United Kingdom, and is in the midst of comprehensive planning in the Netherlands. The scan team traveled to Europe and Singapore to meet with transportation officials with expertise in road pricing programs to learn firsthand about their approaches and practices.

While a number of basic objectives may underlie a road pricing program, the scan team focused on two primary purposes of road pricing: to manage demand and to generate revenue. Figure 1 illustrates the fact that some programs emphasize one objective, and others seek to blend the two objectives in one harmonious program. The countries visited in this scan can be viewed through the lenses of revenue generation and demand management. London, Singapore, and Stockholm are in the demand management circle, while the Czech Republic and Germany fall solidly in the revenue generation circle. The aim of the Netherlands is to transition on a revenue-neutral basis to a more fair transportation funding system that charges users for vehicle use instead of vehicle ownership. A clear understanding of the primary policy objectives behind the implementation of road pricing and consistent decision-making aligned with the objectives were essential elements for all successful projects reviewed as part of this scan.

Both the U.S. Department of Transportation (U.S. DOT) and the American Association of State Highway and Transportation Officials (AASHTO) have made economic and environmental sustainability and community livability top priorities. The evidence in the countries visited in this scan shows that road pricing can play a vital role in creating new funding for transportation, encouraging improved quality of life in the urban environment, advancing economic productivity for goods movement and business, increasing the use of public transit, and reducing congestion and emissions.

Major Findings

Over a 12-day period, the scan team interacted with host country experts to develop an understanding of the political, institutional, and technical factors that contributed to the successful implementation of road pricing. Based on discussions and observations made during and after the scan, the team developed the following nine major findings:

1. Host countries and regions with **clearly defined and well-understood policy goals** were able to achieve their targeted outcomes most effectively.
2. A **large-scale demonstration** project is a powerful tool for building public acceptance, allowing people to experience the benefits of congestion pricing.
3. Thorough **planning and performance measurement** pay benefits in ensuring achievement of overall goals, managing the pricing program as an element of overall transportation system performance, and directing implementation and operations effectively.

4. **Linking the pricing structure to the benefits received** by the user contributes to public acceptance and helps avoid the potential negative impacts of traffic diversion.
5. **Public outreach and communications** were key components of the program at every stage: before making the implementation decision, during the program design process, and during the operational phase.
6. **Open-source system designs offer long-term advantages** by leveraging market competition to manage implementation and operations costs, ensure system flexibility and scalability, and establish a foundation for system interoperability.
7. **Interoperability** among states and countries is recognized as a critical issue that needs to be addressed at high levels.
8. **Equity and privacy** concerns are addressed by host countries through exemptions, revenue use, technology, and business rules.
9. The urban area pricing projects integrated **public transit investments and land use planning** to manage congestion.

These findings are complemented by more than 30 documented lessons learned that are characterized by seven functional categories and business disciplines. The lessons learned are intended to provide more in-depth discussions of the findings from discussion with the host countries. They are organized to enable subject matter experts to examine the areas of greatest interest most effectively.

Implementation

Based on the key findings and lessons learned, the team recommends that additional resources and effort be focused on three strategic areas:

1. **Enhanced outreach and communications.** To advance the use of road pricing in the United States, it is paramount that transportation leaders, policymakers, key stakeholders, and a larger cross-section of the public understand the benefits and implications of broader road pricing.
2. **Additional research needs.** With the application of road pricing in the United States limited to high- occupancy toll lanes, there is a continued need for additional research to better comprehend issues related to public perception, implementation barriers, behavioral effect, and integration of road pricing with multimodal land use and transit options.
3. **Road pricing toolkit.** Transportation professionals lack a comprehensive decision analysis tool to assess the merits of various road pricing options to address specific problems. The toolkit would include a module to assist in making design decisions, a guidebook or primer to assist technical managers in developing financing and procurement strategies, comprehensive and synergistic transportation plans that incorporate road pricing concepts applicable in the U.S. context, and analytical tools to estimate performance and costs of alternative concepts in comparison with conventional tax-based approaches. The tools would culminate in a decision tree to help transportation leaders make informed decisions on the relevance and feasibility of a road pricing alternative.

1. Introduction

Transportation networks affect virtually every aspect of peoples' lives—where they live, where they work, where they shop, and how much they pay for goods and services. In many U.S. metropolitan areas, congestion is a significant and growing problem. Road pricing holds great promise as a strategy for reducing the absolute levels of congestion, while at the same time raising revenue for transportation, reducing environmental impacts of traffic, and allowing people to consider the full cost of transportation and land use decisions. The primary purposes of road pricing include congestion or value pricing to manage demand, environmental pricing to reduce environmental impacts, and tolling to generate revenues (see "Types of Road Pricing").

The scale of application of road pricing may also be used to categorize it into facility-based pricing, zone-based pricing, or distance-based pricing (see "Scale of Application").

At any scale of application, charging systems may have rates that vary by time of day, traffic volumes, vehicle classification, and other factors to garner the advantages of a congestion pricing approach or to address environmental impacts.

U.S. transportation agencies have a long history of facility-based tolling for revenue generation and have made great strides in managing demand on specific highway lanes using congestion pricing. However, broader scale applications of road pricing (i.e., zone-based or regionwide) have been limited because of political, institutional, and public acceptance issues. Zone- based pricing (e.g., in urban centers) and regionwide pricing of roadway systems have fallen short of implementation in the United States and have been evident only in research, demonstration pilots, and proposals that failed to get to implementation, such as the New York City cordon pricing proposal.

Variable charges have been used successfully to manage demand in many U.S. industries, including the pricing of hotels, airfare, electricity, and mobile phone services. Many economists view road pricing, or using pricing signals to manage demand, as the single most viable approach to reducing traffic congestion. Yet, the application of variable pricing in the United States has been limited primarily to a few toll facilities and high-occupancy toll (HOT) lanes in a handful of cities.

Both the U.S. Department of Transportation (U.S. DOT) and American Association of State Highway and Transportation Officials (AASHTO) address traffic congestion in their programs. U.S. DOT's Urban Partnership (UP) and Congestion Reduction Demonstration (CRD) programs have set the stage for adding several facilities to the list of U.S. examples of pricing. One of the UP/CRD cities, Seattle, WA, will price all lanes on an existing facility. In addition, U.S. DOT and the Federal Highway Administration (FHWA) provide technical assistance and conduct research on road pricing as a demand-management and revenue-generation strategy. FHWA established an office to promote implementation of innovative program delivery strategies, such as road pricing and public-private partnerships (PPP).

One of AASHTO's key objectives under its strategic goal to reestablish transportation as a national priority recognizes the need to increase mobility by encouraging multimodal and intermodal solutions, policies, and technologies. AASHTO's goal to "provide world-class technical services" includes facilitating the use of emerging technologies, processes, and programs and advancing innovative practices. Road pricing is an innovative strategy that can help achieve these goals.

Types of Road Pricing (Based on Purpose)

- **Congestion or value pricing**—Fee charged by time of day or traffic volume to reduce traffic, manage traffic flow, and/ or maintain target travel speeds.
- **Environmental pricing**— Fee charged to reduce the environmental impacts of traffic.
- **Tolling**—Generic fee paid to access a road, bridge, or tunnel to generate revenue to cover the costs of construction and/or operations.

Scale of Application of Road Pricing

- **Facility-based charges**— Charges for use of specific roadway facilities or dedicated lanes on these facilities.
- **Zone-based (area or cordon) charges**—Variable or fixed charges to drive within or into a congested urban area. Under a cordon system, such as in Stockholm and Singapore, users pay a fee every time they cross the cordon boundary. In an area-priced system, such as in London, users pay a fee to enter, leave, or travel within a defined area.
- **Regionwide or nationwide distance-based charges**—Per-mile charges on all roads within a prescribed region or country.

The results of this international road pricing scan will inform the U.S. road pricing research agenda, but, more important, it will identify best practices from international experience to assist U.S. practitioners in considering and implementing broader road pricing strategies.

Purpose of Road Pricing Scan

The purpose of the scan was to identify new ideas and practical, workable models for integrating road pricing approaches into State, local, and regional policies, programs, and practices. This scan reviewed urban and nationwide road pricing approaches in Europe and Singapore so that the U.S. participants could develop an understanding of the political, institutional, and technical factors that contributed to their successful implementation and, in some cases, their rejection. These insights have helped the scan participants recognize the conditions and objectives in which road pricing can play a productive and meaningful role. Their perspectives will be communicated to a broad U.S. audience of policy- and decision makers.

The best practices from the places visited will be used to develop and apply new strategies for implementing broader forms of road pricing in the United States and help focus attention on the potential for road pricing as an effective part of 21st century transportation operations and financing policies, programs, and practices.

Panel Scope, Sponsorship, and Composition

The U.S. panel met with officials from Berlin, Germany; the Czech Republic; London, United Kingdom; Singapore; Stockholm, Sweden; and The Hague, Netherlands, from Dec. 7 to 18, 2009. The face-to-face visits enabled participants to gain a deeper understanding of each host country's history and context, the goals and objectives that were established, how road pricing was designed to address transportation and policy objectives, and the hurdles that were faced and how they were overcome. The exchanges provided an opportunity to gain in-depth understanding of program goals and methods, implementation costs, benefits, transportation impacts, revenue generation and use, operating and technical practices and their costs, financing approaches, effects on safety and the environment, and public acceptance.

The panel was cosponsored by AASHTO, FHWA, and the National Cooperative Highway Research Program (NHCRP). The 10 members of the multidisciplinary team included transportation professionals from four State departments of transportation (DOT), one regional transportation agency, FHWA, the Federal Trade Administration (FTA), and private industry. Bob Arnold of FHWA and Vance Smith of the Georgia DOT were the cochairs, Patrick DeCo rla-Souza of FHWA served as the Implementation Team chair, and John Doan of SRF Consulting Group was the report facilitator. Other team members were Rodney Barry of FHWA, Jayme Blakesley of FTA, Mark Muriello of the Port Authority of New York and New Jersey, Gummada Murthy of the Virginia DOT, Patty Rubstello of the Washington State DOT, and Nick Thompson of the Minnesota DOT. Contact information and biographies for each team member are in Appendix A. Details on the scan preparation and itinerary are in Appendix B.

Panel Topics of Interest

Major topics of interest included the following:

- Urban and nationwide pricing
- Strategies for addressing political, regulatory, and legal barriers, particularly those related to public acceptance
- Institutional arrangements and interagency collaboration to enable effective applications of pricing techniques
- Implementation strategies and costs
- Experience with quantifying projected and actual benefits (e.g., congestion reduction, safety, and environmental) and developing performance metrics
- Relationship between road pricing as a revenue stream and operational strategy to reduce congestion, improve safety, or enhance the environment
- Equity concerns, particularly redistribution issues and strategies related to toll revenues
- Relevance and importance of including supporting strategies such as transit enhancements, system operations strategies and technologies, and travel demand management

Figure 2. Scan team members in Singapore (left to right): Patrick DeCorla-Souza, Rodney Barry, Nick Thompson, Patty Rubstello, Jayme Blakesley, John Doan, Mark Muriello, Bob Arnold, Gummada Murthy, and Vance Smith.

Specific questions on the panel's interests in these topics are in Appendix C. These amplifying questions were sent to country hosts before the U.S. visits to help them determine whom to invite to the meetings with the U.S. contingent and to ensure that presentations addressed the interests and needs of the scan team.

Host Delegations

During the scanning study, the team members met with representatives from more than a dozen organizations that represented a broad range of road pricing practitioners, executives, and key stakeholders. The majority of the organizations represented one of the following perspectives: road agency (city, regional, or national), infrastructure financing agency, trade group, or research organization. In addition, the scan team had informal conversations with residents of the locations visited, including taxi drivers, hotel staff, tour guides, and others, to better understand the local context.

2. General Observations

Road pricing projects have been implemented on five continents (Asia, Australia, Europe, and North and South America) and include notable facilities in Canada, Chile, Germany, Singapore, Sweden, and the United Kingdom. Bhatt et al. provided a concise overview in *Lessons Learned from International Experience in Congestion Pricing* (2008):

> Over the past 30 years, congestion pricing concepts have received considerable attention outside of the U.S. Compared to the U.S., Britain, Europe, and countries in Asia and the Pacific region have a longer history of interest in exploring the potential of pricing approaches to address congestion, environmental and transportation funding problems. Individual countries, as well as the European Union (EU), have established road pricing initiatives aimed at studying, implementing and evaluating a wide range of congestion pricing demonstrations and operational programs.

These projects have demonstrated that road pricing can be an effective means of managing demand and generating revenue and can also be politically and publicly acceptable. Pricing programs have reduced congestion on facilities and in priced areas, improved use of existing road capacity, created new travel options to driving adopted by travelers, and achieved the goals of demand management, emission reductions, and revenue generation. Revenues from pricing have been used to provide funding for multimodal transportation improvements. Much like the U.S. experience, overseas road pricing projects have been met with considerable resistance and political and public debate. Models of new roadway capacity, propelled by the toll financing that is more prevalent overseas, are being met with less resistance and often allow for private sector participation in construction, finance, and operation of newly built priced roads. While models involving distance-based highway charges for commercial vehicles and urban zone-based congestion pricing have encountered far greater scrutiny and debate, they have tended to achieve more targeted transportation goals. They also have achieved broader objectives aimed at addressing the environmental, livability, and quality of life impacts of transportation in urban areas. Many international examples indicate that public acceptance and approval of pricing programs improves significantly after project implementation, when the benefits and impacts can be weighed in tangible terms based on the context of its application.

Many international road pricing programs are breaking new ground and providing important lessons for those interested in exploring the use of market-based approaches to address traffic congestion. Projects implemented to date have proven that travelers are willing to pay for road use provided there are demonstrable improvements to travel conditions coupled with other enhancements to the transport system. Pricing signals have had measureable effects on the traveling public's transportation decisions, evident in more efficient use of existing roadways and other transportation facilities, such as transit services. Although road pricing is operational in multiple locations abroad, it is still a new and innovative concept in the United States—one that requires careful planning, coalition building, public education and participation, and sufficient time and resources to develop well-designed and locally acceptable project plans.

Table 1 (see next page) summarizes the characteristics of the road pricing projects the team visited during the scan. Based on similar purposes and characteristics, London, Singapore, and Stockholm use urban congestion pricing concepts that aim to reduce urban congestion by **managing demand** through zone-based (i.e., area or cordon) or facility-based (as on Singapore expressways) pricing. The Czech Republic and Germany have distance-based charging systems for heavy commercial vehicles on selected highways with the primary goal of **generating revenue.** The Netherlands is unique among the sites visited because it is in the planning and implementation stages for a nationwide road pricing system that may charge all vehicles based on distance traveled, time of day, and vehicle type for the **combined purposes** of managing demand and generating revenue to shift to a more user pays basis.

Each host country has unique geographic, cultural, and political contexts that drive the decision-making and public involvement process. All of the countries have parliamentary democracy forms of governance, in which the executive power lies in a cabinet led by a prime minister and supported by parliamentarians who lead the ministry of transport. In addition, the urban pricing regions visited (London, Singapore, and Stockholm) have high-density development, making certain types of road pricing more applicable and the provision of multimodal services such as transit, biking, and walking more feasible and effective.

Table 1. Summary of pricing projects

	Countries Visited	Purpose/ Objective	Type of Pricing	Milestone Dates	Technology	Measured Impacts	Annual Revenues and Cost (in USD)*	Distribution of Net Revenues
Demand Management	**Stockholm, Sweden:** Congestion Tax	Manage congestion (primary) Promote transit and protect environment (secondary)	Cordon pricing in city center by time of day at SEK10 to SEK20 (about US$1.50 to US$3) per crossing of cordon line into and out of city center	Trial: January– July 2006 Referendum: September 2006 Permanently reinstated: August 2007	Automated number plate recognition (ANPR) to assess tax to vehicle owner	20% reduction in traffic congestion in the city center 10–14% decrease in emissions	Gross revenues (2009): SEK850 million (US$118.5 million) Net revenues (2009): SEK530 million (US$74 million) Overhead costs: SEK320 million (US$44.5 million), about 37% of revenues	Collected by national government and transferred to the city of Stockholm Net revenues used to invest in transit and new roads Net revenues used to invest in transit and new roads
	London, United Kingdom: Congestion Charge	Manage congestion (primary) Promote transit and protect environment (secondary)	Area pricing in central London and its western extension Flat daily rate of £8 (US$13)	Started in central London: February 2003 Price increased from £5 to £8 (60% increase)in July 2005 Western extension: February 2007 Repeal of western extension: planned in 2010	ANPR to track compulsory payment compliance and identify violators	Initial traffic reductions of 25% and 19% (central London and western extension, respectively)	Gross revenues (2008): £268 million (US$435 million) Net revenues: £137 million (US$222 million) Overhead costs: £131 million (US$212 million), about 50% of revenues Capital costs for central London zone: £130 million (US$211million)	Net revenues used for transit (80%) and other transport (20%) improvements within greater London
Revenue Generation	**Singapore:** Electronic Road Pricing (ERP)	Manage congestion (primary) Promote transit (secondary)	Cordon and express-way pricing by time of day and vehicle class	Cordon pricing via manually enforced paper permit system in 1975 Transition to ERP in 1998, followed by expressway pricing	Dedicated short-range communications (DSRC) in-vehicle units with removable stored-value smart card for payment ANPR for enforcement	Achieves free-flow road speed targets of 45–65 km/h on express-ways and 20–30 km/h on arterials.	Gross revenues (2008): SG$125 million (US$90 million) Net revenues: SG$100 million (US$72 million) Overhead costs: SG$25 million (US$18 million), 20% of gross revenues	Net revenues returned to vehicle owners through tax rebates—heavy investment from general fund in transit and highway systems
	Germany: Heavy Goods Vehicle (HGV)	Generate revenue and promote user-pays principle (primary)	Truck tolls for HGVs greater than12 metric tons on the autobahn and limited	Opened in January 2005	Global Positioning System (GPS) for vehicle location	Violations less than 2% Empty truck trips declined by 7% 58% shift from	Gross revenue (2008): €3.5 billion (US$5 billion) Overhead costs: 15–20% of gross revenues Average toll rate: €0.163	Net revenues for roads (50%), rail (38%), and waterways (12%) €560 million (US$815 million)

	Countries Visited	Purpose/ Objective	Type of Pricing	Milestone Dates	Technology	Measured Impacts	Annual Revenues and Cost (in USD)*	Distribution of Net Revenues
	Charging on Highways	Protect environment and encourage mode shift to rail and water (secondary)	portions of other national highways based on distance traveled, number of axles, and emissions class		Global System for Mobile Communications (GSM) for data transmission DSRC and ANPR for enforcement Manual booking system via kiosk terminals and Internet for those without onboard units	dirtier truck models (Euro class 1, 2, 3) to cleaner trucks (Euro class 4, 5)	per km (US$0.378 per mi)	per year for trucker "harmonization" program
	Czech Republic: Truck Charging on Highways	Generate revenue and promote user-pays principle (primary) Advance environmental objectives (secondary)	Truck charges on selected national highways based on distance traveled, number of axles, and emissions class	Opening: January 2007 Originally for HGVs >12 metric tons Expansion to include trucks >3.5 metric tons in January 2010	Transponder-based DSRC system with gantries on mainline highways ANPR for enforcement	Average toll rate of US$0.35 per mi on freeways	Gross revenue (2008): CZK6 billion (US$340 million) Overhead costs: 30% of gross revenues Average toll rate: CZK4.05 per km (US$0.36 per mi) for highways; CZK1.90 per km (US$0.17 per mi) for first-class roads	Net revenues for roads and highways, railway lines, and inland transport routes
Planned	**The Netherlands**: National Distance-Based Tax	Planned to manage congestion, replace vehicle tax revenue, and promote user-pays principle (primary) Promote transit and protect environment (secondary)	National distance-based road pricing of all vehicles (commercial trucks and private cars) on all roadways	Phased implementation originally planned to begin in 2011, with all trucks covered by 2012 and all vehicles by 2018 Implementation on hold because of parliamentary elections in June 2010	Under development, likely GPS for vehicle location, GSM-based data communication, and DSRC interrogation with ANPR for enforcement	2020 forecasted results: 10–5% reduction in vehicle-miles traveled 40–60% reduction in delays 10% reduction in CO_2 6% increase in public transit use	Gross revenues (2019 forecasted): €9 billion (US$13.1 billion) Overhead costs: to be determined (capped in law at 5% of gross revenues) Capital costs (estimated): €5.7 billion (US$8.3 billion)	Revenues intended to replace existing vehicle ownership taxes

*See Appendix F for currency exchange rates.

Demand Management: Stockholm's Congestion Tax

The purpose of the Stockholm road pricing project is to reduce traffic congestion and vehicle emissions in the inner-city area. It was initially introduced by the Green Party and Social Democrats as a full-scale trial with the objectives of reducing congestion, improving access and mobility, promoting transit, and promoting environmental sustainability.

Stockholm's Congestion Tax (Trãngselskatt I Stockholm)

Purpose: Manage congestion (primary), promote transit and reduce emissions (secondary)
Application: Urban cordon pricing
Fee structure: Time-of-day pricing
Use of revenues: Fund transportation and transit improvements in the city of Stockholm
Technology: Automated number plate recognition (AN PR)
Milestones: January 2006 (trial began), July 2006 (trial ended), September 2006 (referendum), August 2007 (reopened)
Managing authority: Swedish Transport Administration
Other highlights: Coupled with significant transit investment in equipment, facilities, and operations; exempts through traffic (30 percent of vehicles from Lidingö Island and on Essingeleden bypass); 7-month preimplementation demonstration
Results: 20 percent reduction in traffic, 10 to 14 percent decrease in emissions, 2 to 10 percent improvement in air quality

The concept involves charging a variable tax for crossing a cordon drawn around the Stockholm city center. Vehicles registered in Sweden are charged when they pass one of 18 control points while entering or exiting the congestion zone in the city's center on weekdays between 6:30 a.m. and 6:30 p.m. (figure 3). The rates vary from 10 Swedish kroner (SEK), or about US$1.50, during off peak hours to SEK15 (about US$2.25) during the shoulders of the peak periods and SEK20 (about US$3) during the peak periods (7:30 to 8:30 a.m. and 4 to 5:30 p.m.). Vehicles are charged each time they cross a control point, up to a maximum of SEK60 (about US$9) per day (figure 4).

Because the congestion charge is managed as a tax by the national government, any change in the fee schedule requires parliamentary action. It has not yet been determined how this provision for price adjustments will impact the effectiveness of maintaining roadway conditions at uncongested levels. No price change has been contemplated since the system was permanently adopted in August 2007.

The plan was originally developed in spring 2003. The options for the charging zone were carefully analyzed using traffic simulation tools that defined the details of the system. The traffic analysis was a critical element of proper planning, ensuring that there would be no unintended impacts elsewhere in the network as a consequence of the congestion charge. The planners originally contemplated more complicated charging schemes, but the program's schedule constraints did not allow these to be pursued. The final system was simple and easy to understand, which contributed to successful operation and public acceptance.

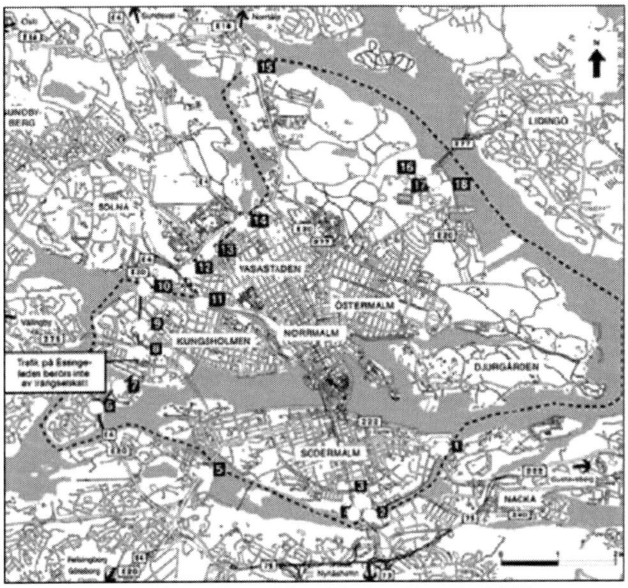

Source: Gunnar Soderholm Presentation, City of Stockholm, Dec. 8, 2009.

Figure 3. Eighteen congestion charging control points create the Stockholm cordon.

Congestion charges and times

Peak Periods		
7.30-8.30 a.m., 4-5.30 p.m.	SEK20	€2
Semipeak Periods		
7-7.30 a.m., 8.30-9 a.m. 3.30-4 p.m., 5.30-6 p.m.	SEK15	€1,5
Medium-Volume Periods		
6.30- a.m., 9 a.m.-3.30 p.m. 6-6.30 p.m.	SEK10	€1
Maximum Charge:	SEK60/day	€6
Evenings, Saturdays, Sundays, holidays: **No Charge**		

Figure 4. Congestion charge fee schedule.

The road pricing project in Stockholm was unique in that it was introduced with a 7-month demonstration period, after which the system was turned off and subjected to a public referendum by Stockholm residents to inform a final decision by policymakers on its continuation as a permanent system. In June 2003 the Stockholm City Council voted to conduct a trial implementation of congestion pricing in its central business district (CBD). The project became operational in January 2006 and was shut down as scheduled at the end of July 2006. Six weeks later, on Sept. 17, 2006, Stockholm residents voted on whether to reactivate or terminate the road pricing system.

Before the demonstration, public support for the pricing program was at 25 percent. After the demonstration, public support from Stockholm residents voting in the referendum was more than 50 percent in favor of reinstating the congestion tax. The referendum counted only Stockholm residents who realized the most tangible benefits of congestion pricing coupled with significant investments in transit. Optional votes taken in some of the surrounding suburbs at the same time as the city referendum showed less than 50 percent support from those communities. These votes were not mandated by Parliament, but were requested by localities that questioned the wisdom of the scheme. Recent opinion polls show that 65 percent of the public would vote in favor of the system in its current form, 17 percent would like to see the price raised, and 25 percent would like to see the price decreased (figure 5, see next page).

The overall implementation included a SEK1.3 billion (US$180 million) investment for the tolling system plus SEK2 billion (US$280 million) in related public transit improvements. The transit investment funded a 10 percent expansion of the Stockholm public transport system, which included 200 articulated buses (equivalent of 10,000 new seats), 2,400 new park-and-ride spaces, bus priority at traffic signals, improved rail service, new dedicated bus lanes, and 12 new express bus routes (figure 6, see next page).

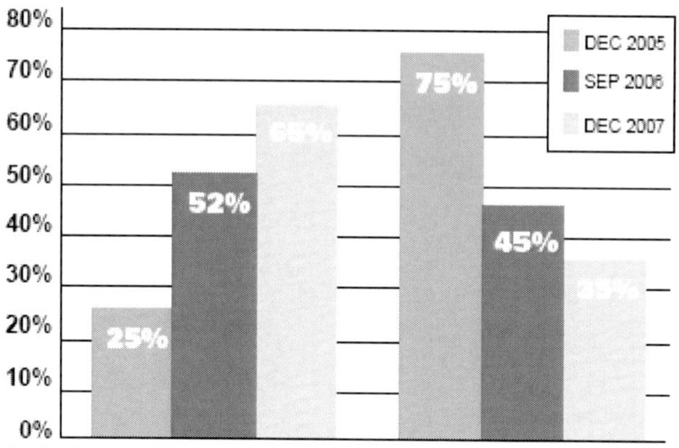

Source: Gunnar Soderholm Presentation, City of Stockholm, Dec. 8, 2009.

Figure 5. Public support for Stockholm congestion tax (2005–2007).

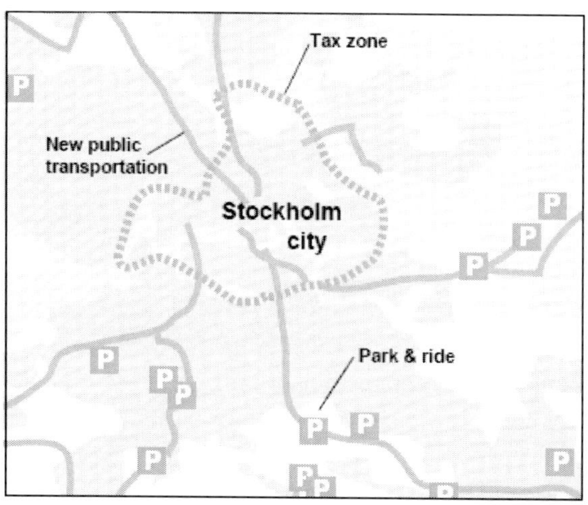

Source: Bhatt, et al. "Lessons Learned from International Experience in Congestion Pricing."

Figure 6. Stockholm pricing zone, new transit routes, and expanded park-and-ride facilities.

Originally, the transit and tolling systems were scheduled to be launched in tandem. However, the transit service enhancements went into operation 5 months before the congestion charging system because of tolling system procurement delays. Transit usage and traffic congestion levels in the city center did not change with the introduction of the new transit service. It was only after the congestion tax was implemented that transit use grew significantly and a 20 percent reduction in traffic was realized. Half of the operation and maintenance cost associated with the new bus services was covered by fares and half by taxes.

Vehicles exempt from paying fees include public buses, taxis, certain alternative-fuel vehicles, emergency vehicles, motorcycles, vehicles with handicap plates, and foreign-registered vehicles. An exemption is also provided to residents of the Island of Lidingö, who can access the rest of Sweden only by traversing the CBD. Vehicles driving between the island and the bridge control points have 30 minutes to make a through trip without being charged. However, if they remain in the CBD for more than 30 minutes, charges are applied. The capital cost to institute the Lidingö Rule was estimated at more than SEK200 million (US$28 million), plus ongoing exemption costs from system downtimes. For example, if even a single gantry is not functioning, all charges must be suspended on the entire system to avoid the possibility of an erroneous charge for an exempt through trip from Lidingö. Another exception is through traffic on the E4/E20 Essingeleden Highway going past Stockholm. In total, about 30 percent of vehicles passing through the city center are exempt from the congestion tax.

The toll collection system includes three overhead gantries at each control point (figure 7). The outer two gantries house the digital imaging cameras that capture front and rear license plate images of all vehicles. The middle gantry houses the dedicated short-range communications (DSRC) antennas used in conjunction with in-vehicle transponders available to travelers during the trial. With the adoption of the permanent system, officials decided that the automated number plate recognition (AN PR) system performed so well that the transponder-based option was not necessary, and eliminating it offered an opportunity to reduce overall system operating costs.

Source: Gunnar Soderholm Presentation, City of Stockholm, Dec. 8, 2009.

Figure 7. One of 18 control points using the three-gantry electronic toll collection system.

During the congestion tax trial, vehicles with transponder- based accounts and those that did not have transponders had until noon the following day to pay the charge via the Internet or at a physical retail location, which included 7-Eleven and Pressbyran convenience stores, Giro Banks, and stand-alone kiosks. Charges not paid during the grace period are assessed a SEK70 late fee (US$8) for the first reminder and SEK450 (US$62) fee for the second reminder. Swedish laws governing the collection of taxes required that the system capture all license plates, whether or not a transponder was used. Therefore, the transponder-based system was duplicative. Because of the extra operating costs resulting from administration and complexity of the transponder-based payment option, it was phased out with the implementation of the permanent system. The options for payment were also changed to increase convenience for taxpayers. Under the current arrangement, drivers receive a monthly billing statement from the state.

The Stockholm system processes about 450,000 transactions per day. The original IBM contract was worth SEK1.9 billion (US$265 million), and was contracted through the Swedish Road Administration. The gross revenue from the system in 2009 was SEK850 million (US$118.5 million). The cost of 2009 operations was about SEK320 million (US$44.5 million), or about 38 percent of total revenues. With anticipated process streamlining and operating cost reductions, operation costs were expected to decrease to SEK220 million (US$30.6 million) in 2010 and 2011.

The operating cost reduction initiative for the system is a serious multiyear effort with clear established targets. The fear of failure for the technology loomed large as the system was being designed for the trial period. The Stockholm system planners knew that the trial system needed to run nearly flawlessly to gain public confidence and acceptance. The consequence was an expensive system to operate. This has led to an aggressive refinement of operating practices to reduce ongoing costs.

Overall, the congestion tax reduced traffic volumes by 10 to 15 percent, and congestion fell by 20 percent in the CBD. Transit ridership grew by 6 to 9 percent during the demonstration. This resulted in a 14 percent reduction in vehicle-miles traveled and a 10 to 14 percent decrease in emissions. After the system was taken offline at the end of the demonstration, traffic volumes returned to about the same level as before the trial. After its permanent reintroduction in August 2007, the data show that access to the CBD improved, travel times were lower, and a 20 percent reduction in congestion was realized.

Demand Management: London's Congestion Charge

In 2003, under the leadership of Mayor Ken Livingstone, London launched a bold initiative to designate a congestion charging zone in central London and charge vehicles to travel within the 8-square-mile (20.7-square-kilometer) area (figure 8, see next page). In 2007, the charging area was doubled in size with a western extension. Because of changes in leadership and public consultation, the western extension was scheduled to be repealed in 2010.

With London roads congested most of the day before the congestion charge, it was estimated that 40 percent of England's congestion was in greater London, with central London the most congested. Average all-day speeds were less than 9 miles per hour (mi/h) (14.4 kilometers per hour (km/h)) in central London. Delays were costing people and businesses £4 million to £6 million (US$7 million to US$10 million) per week in time and money. Thus, the objective of the congestion charge was to reduce traffic, improve travel times for buses, generate new revenues for public transit, and enhance the quality of life in central London. The mayor's vision included a 40 percent increase in public transit service by 2011, with the immediate expansion of bus services before the start of congestion charging in February 2003.

The flat weekday charge was set at £5 (US$10.50) initially and raised to £8 (US$13) in August 2005. The charge is in effect on weekdays from 7 a.m. to 6 p.m. (figure 9, see next page). Various exemptions and discounts are allowed, including a 90 percent discount for residents living in the pricing zone. Buses, taxis, emergency vehicles, hybrid cars, and motorcycles are exempt as well.

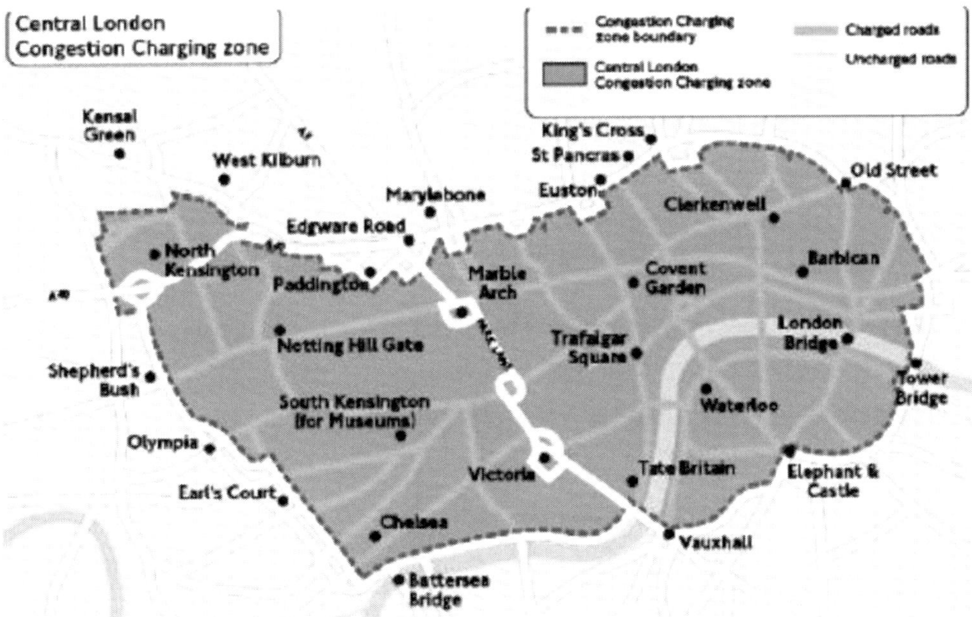

Figure 8. Central London congestion charging zone.

Figure 9. Congestion charging sign.

London's Congestion Charge

Purpose: Manage congestion (primary), promote transit and reduce emissions (secondary)
Application: Urban area pricing
Fee structure: Flat rate of £8 (US$12) per day (90 percent discount for residents in the zone)
Use of revenues: Fund improvements to London transit (80 percent) and transportation (20 percent)
Technology: ANPR
Milestones: February 2003 (central London zone began operations), August 2005 (charge increased), February 2007 (western extension of zone), 2010 (repeal of western extension)
Managing authority: Transport for London
Other highlights: Significant investment in transit, repeal of western extension in 2010, 30 percent of vehicles entering the zone are exempt
Results: 25 percent and 19 percent reduction in traffic (central and western extension, respectively), net annual revenues of £137 million (US$220 million) in 2008

Enforcement is conducted using 1,360 closed-circuit cameras at 348 sites along the cordon and within the zone with roving vans policing the zone, capturing live video images of the license plates of all vehicles (see figure 10). Every operational weekday, the ANPR system captures and processes 1.45 million images. The daily charge must be paid by the end of the day following the day the charge was incurred and can be paid online, by phone or text message, or in specific stores with the "epay" or "congestion charge" logo. If the charge is not paid by midnight of the day following the day of the charge, the cost increases by 25 percent to £10 (US$16.25). The penalty for nonpayment is about £40 (US$65) if not paid until after mail receipt of a penalty charge notice (figure 11). To minimize erroneous penalties, key data and images for every possible penalty charge is manually checked before being issued. The penalty triples to £120 (US$195) if not paid within 4 weeks. Users can pay up to 90 days in advance for travel within the charging zone. In addition, monthly and annual passes are available at a discounted rate.

Figure 10. A total of 1,360 cameras are located on all routes into and out of the zone as well as within the zone.

Congestion charging

Transport for London

Penalty Charge Notice

Penalty charge payment slip

MAYOR OF LONDON

Figure 11. Example of a penalty charge notice.

After implementation of the London congestion charge, the number of vehicles (four or more wheels) entering the charging zone decreased by 25 percent, or 70,000 fewer vehicles per day, and has remained constant. The amount of circulating traffic fell by 15 percent after the first year of implementation. Travel speeds increased by 30 percent, trip times decreased by 14 percent, and traffic delays plummeted by 25 percent in the charging zone. Transport for London (TfL) reported an average of 70,000 fewer daily vehicle trips than in the year before the congestion charge. Of those reduced trips, an estimated 50 to 60 percent shifted to transit (figure 12), 20 to 30 percent of the trips were eliminated, and 15 to 25 percent involved carpooling.

Figure 12. Public service advertisement promoting shift to transit.

Despite the successes of the London congestion charge, traffic congestion in central London has gradually returned. In 2009, congestion levels within the central charging zone were equal to those evident before 2003. This is partly because of the decision to take advantage of the lower vehicular use of London's central roadways by converting portions of various streets or entire streets for pedestrians and bicycles and for dedicated bus lanes. The reconstruction of urban infrastructure (e.g., water and sewer lines) and construction related to preparations for the 2012 Summer Olympics have also consumed some street capacity, adding to vehicle congestion levels.

Revenues from the congestion charge were £268 million (US$435 million) in 2008. When accounting for expenses (about 50 percent), the congestion charge generated about £137 million (US$222 million) in the same year, which by law must be spent on transportation in greater London. Of the 2008 net revenues, 82 percent went for bus improvements, 9 percent for roads and bridges, and the remaining 9 percent for road safety, pedestrian and cycling facilities, borough plans, and environmental improvements (figure 13).

While the objectives of the congestion pricing program in London were achieved, critics point to the high operating costs of the system as a drawback. Much of the success of the system from a user's perspective rests with its relative simplicity, with only one price (£8 per day) charged when the system is in operation. Based on TfL's 2006 and 2007 reports, the business and economic impacts of congestion pricing have been largely neutral. Furthermore, a 2007 Coordination of Urban Road-User Charging Organizational Issues (CURACAO) report stated, "The level of acceptability of road user charging before the introduction was rather stable at about 40 percent. This also holds true in comparison with other scenarios such as workplace commuter tax schemes. After the introduction acceptability has risen above 50 percent."

Figure 13. Public service advertisement promoting use of 2008 congestion charge revenues.

Other urban area pricing proposals have been pursued unsuccessfully in the United Kingdom, notably in Edinburgh, Scotland, and Manchester, England. In both cases, the proposals were put to voter referendums before implementation. Without the opportunity for constituents to experience the tangible benefits of the road pricing initiatives, both proposals failed to garner sufficient voter support at the polls. Nationally, because of the 2010 elections and perceived lack of public support, neither major party actively considered or embraced road pricing on a broader level at the time of the scan.

Demand Management: Singapore's Electronic Road Pricing

Singapore is distinctive in its geography as an island city-state situated at the southern end of the Malaysian peninsula, which limits the capacity for physical growth. Singapore's politics and culture are supportive of its ability to address these constraints by integrating multimodal transportation policies that incorporate stringent land use and vehicle ownership standards.

In 1975, Singapore undertook a bold new approach to manage traffic and improve air quality by introducing a fee for vehicles entering the CBD during the morning peak period of 7:30 to 9:30 a.m. Since its inception as a nonelectronic, prepaid windshield permit, Singapore's road pricing system has expanded and modernized to become the most extensive congestion pricing system in the world.

Electronic Road Pricing

Purpose: Manage congestion (primary), promote transit (secondary)
Application: Cordon with time-of-day pricing in city center, as well as expressway and outer ring roads
Fee structure: Varies by time of day, location, and vehicle classification, with prices revised quarterly to maintain travel speed standards
Use of revenues: Returned to vehicle owners through tax rebates (through the general fund)
Technology: DSRC with in-vehicle units (known as IUs) equipped to accept smart cards tied to banking for payment, ANPR for enforcement
Milestones: 1975 (original permit scheme opened), 1998 (electronic charging began)
Managing authority: Land Transport Authority
Other highlights: Multimodal investments, vehicle quota system, integration with land use and transit planning, smart card technology protects user privacy, use of interactive gallery exhibit to educate younger audiences about road pricing and other transportation options
Results: Achieved target speeds of 45 to 65 km/h on expressways and 20 to 30 km/h on arterials, net revenue of SG$100 (US$75 million) in 2008

The paper-based road pricing system was converted to fully electronic in 1998 with the adoption of standardized in- vehicle units that employ DSRC for road pricing. The system also employs ANPR technology for enforcement. Both technologies allow the system to operate with vehicles traveling at full freeway speeds (up to 120 km/h). Known as Electronic Road Pricing (ERP), the system is fully automated, with more than 60 charging points covering the center city and some primary high-volume expressways (see figure 14).

All vehicles registered in Singapore are required to have one of six types of in-vehicle unit (IU) installed (figure 15).

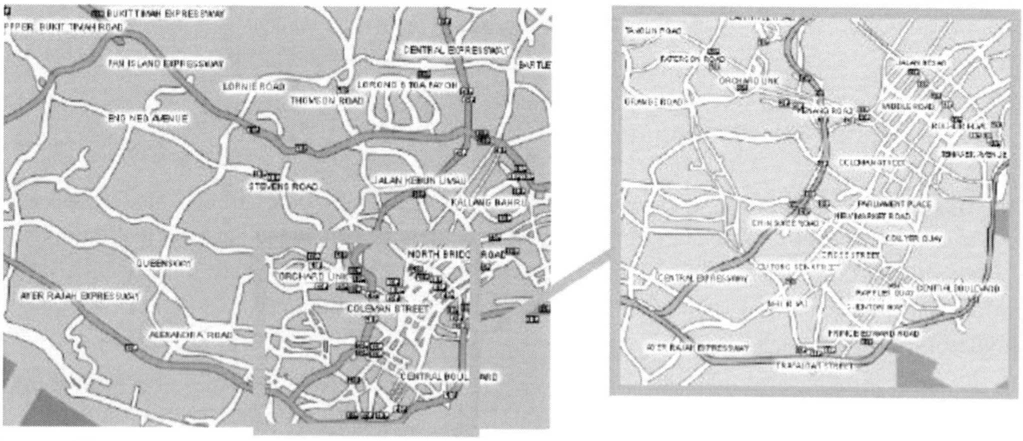

Source: Bhatt et al. Lessons Learned from International Experience in Congestion Pricing.

Figure 14. Singapore's Electronic Road Pricing (2005).

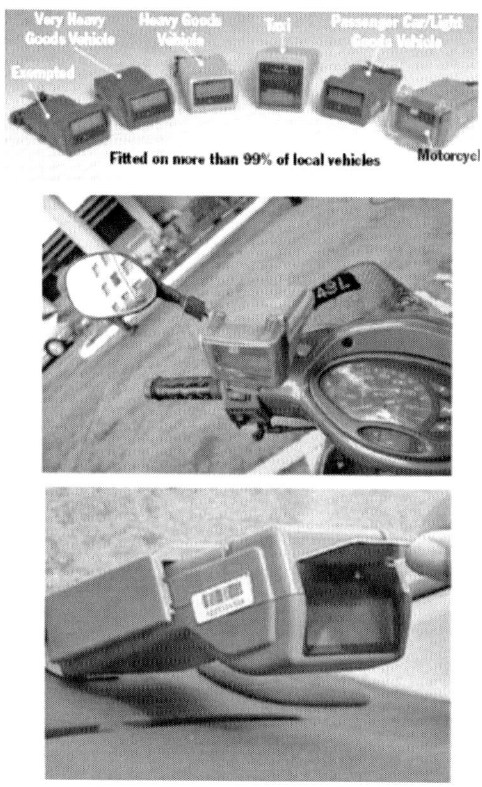

Figure 15. In-vehicle units (six types are available based on vehicle classification) with prepaid stored-value smart cards are required in every motor vehicle, including motorbikes, registered in Singapore.

Each type is programmed to charge different rates based on the vehicle type. For example, a passenger car or taxi is charged for one passenger car unit (PCU), the price (SG$2.00) shown in the upper right of the pricing sign (see figure 16, next page). A motorcycle is charged 0.5 PCU, in this case SG$1.00, while a small bus is charged 1.5 PCU (SG$3.00) and a large bus is charged 2 PCU (SG$4.00).

A separate prepaid, stored-value smart card is inserted into the IU by the driver when the vehicle is in use. The smart card protects user privacy by not storing personal data on the card. The IU has a visual display and audio signal that tells the owner when a toll is charged and when the smart card balance is low. The smart cards are available through various banks and can be replenished at automatic teller machines, kiosks, gas stations, online, and by telephone. A newer generation smart card, called CEPAS (short for Contactless E-Purse Application Standard), was introduced in 2006 with the intent of increasing the card's utility to holders. It can be used to pay for transit fares and parking and at various retail stores.

ANPR technology is used for enforcement. Vehicles without an active transponder detected face a SG$70 fine (US$50), while those with insufficient smart card funds are charged an administrative fee of SG$8 (US$6). This automated enforcement keeps violations at less than one percent. The introduction of ERP was accompanied by new park-and-ride lots, expanded transit service into the CBD, and a 30 percent decrease in CBD parking rates.

Because of the ERP system's longevity in Singapore, its effects on travel behavior and traffic conditions are well known and predictable. By some accounts, it would require a three-fold gas tax increase to cause the same travel behavior change as a one-fold increase in ERP. Overall, prices in the CBD are set to maintain average travel speeds at 25 km/h. In October 2005, the cordon was extended to include the Orchard Area, the region's most prominent shopping district. Traffic and retail sales data were collected before and after ERP was introduced in the Orchard Area. Analysis of these data showed that ERP did not affect destination traffic to the shopping district. Sales continue to be healthy and grow at the same rate as before the ERP system was introduced to the area.

Singapore's ERP system varies its prices by time of day, location, and vehicle classification. Both upward and downward adjustments to the pricing schedule are considered every 3 months based on routine collection of traffic data on roadway speeds. The prices are set to ensure that targeted speed standards are maintained at an 85th percentile level to ensure free-flow speeds for at least 85 percent of vehicles charged. Speed targets are 45 to 65 km/h on expressways and 20 to 30 km/h on other streets. ERP prices can vary for a passenger car from zero to about SG$3 (US$2) per CBD cordon crossing and are in effect from 7 a.m. to 8 p.m. on weekdays. In addition, ERP hours of operations for the Orchard District include Saturdays from 10 a.m. to 8 p.m. On the expressways, the price varies from zero to about SG$5 (US$3.50) and is in effect weekdays from 7 to 9 a.m. and 5:30 to 10:30 p.m. Unlike in Stockholm and London, no vehicles are exempt from the charge in Singapore, except emergency and military vehicles.

A major review of the ERP system was conducted as part of the Land Transport Master Plan in 2008. A number of enhancements were advanced, including 16 new ERP gantry locations, changes to the performance criteria, and rate changes. Traffic analysis showed that these enhancements resulted in an 18 to 25 percent decrease in traffic volumes in the Bugis-Marina area and a 7 to 21 percent decrease in the Orchard District on Saturdays (table 2).

Figure 16. ERP charging point.

Table 2. Summary of traffic results in Orchard and Bugis-Marina areas on Saturdays after July 2008 enhancements

	Traffic Volume Decrease	Traffic Speed	Traffic Speed Improvement
Within the Bugis-Marina Centre Cordon	18–25%	20–22 km/h	3–2%
Within the Orchard Cordon	7–21%	21–25 km/h	1–30%

In 2008, annual gross revenues were SG$125 million (US$90 million), with net revenues of SG$100 million (US$72 million). While all of the net revenues are directed to the general fund, vehicle owners receive periodic rebates on vehicle taxes funded by ERP net revenues. Investments from the general fund are provided to appropriately address the needs of Singapore's multimodal transportation system.

Singapore's demand management techniques involve a combination of vehicle ownership and usage measures. Usage measures include ERP charges, gas taxes, parking restrictions and fees, and off-peak car registration permits that are purchased at a discounted rate and allow use of the vehicle only during weekends and nonpeak hours during weekdays. Car ownership is closely controlled by a quota on the number of vehicle licenses permitted each year. This vehicle quota system was implemented in 1990 to regulate the growth of the vehicle population in Singapore. These permits, called Certificates of Entitlement (COE), are acquired through an open online auction. In November 2009, the cost for a COE ranged from about SG$16,500 (US$11,800) to SG$18,300 (US$13,100). Other costs of vehicle ownership include vehicle registration fees, excise duties, and road taxes. In total, these taxes and fees increase the price of new vehicle acquisition to three to four times the actual cost of the vehicle. For example, a new Toyota Corolla with a dealer cost of SG$17,850 (US$12,800), costs about SG$70,000 (US$50,200) to purchase.

The Singaporean government takes a long-term approach to public outreach and education. It recently opened the Land Transport Authority (LTA) Transportation Gallery, an interactive exhibit that targets school-age children and teaches them about the past, present, and future of transportation planning and policies in Singapore. Gallery admission is free and includes a 1-hour guided tour through the following six transportation exhibits:

- **Journeys**—Provides an overview of land transport systems in the world's major cities, including Singapore's own modes of transportation
- **Memories**—Brings visitors back to experience the transportation modes available during Singapore's preindependence era
- **Formative Years**—Highlights key milestones in Singapore's land transportation development and the challenges faced during the nation's formative years
- **Land Transport Today**—Explains how LTA addresses the challenges of managing road usage and meeting the diverse needs of people
- **Challenge Theatre**—Engages visitors by putting them
- in the role of a transport planner in an interactive, multiplayer game about how to best develop and manage land transportation
- **Vision and Aspirations**—Propels the visitor into the future via interactive multimedia activities that provide a peek at what the Singapore transportation landscape might be like in 2030

The LTA Transportation Gallery is a powerful and engaging tool that teaches the next generation about the need for good transportation planning and how ERP is an effective demand management tool (figure 17).

Revenue Generation: Germany's Heavy Goods Vehicle Tolling

Germany is at the crossroads of Europe, with significant east-west and north-south freight movements via trucks. German motor fuel taxes did not prove to be an effective means of revenue generation to pay for highway infrastructure maintenance resulting from the volumes of heavy trucks moving across Germany. As a consequence, the Germans sought a new source of revenue through distance-based charging on the autobahns, the equivalent of the U.S. interstate highways. The German HGV tolling program began commercial operations in January 2005. It is the world's first satellite-based, countrywide electronic tolling system and applies only to trucks weighing more than 12 tons on the autobahns and a small number of other national highways.

Figure 17. LTA Transportation Gallery interactive exhibit.

Germany's HGV Tolling

Purpose: Generate revenue for transportation funding and shift to user-pays approach (primary), reduce emissions and modal shift to rail/water (secondary)
Application: Distance-based pricing on selected national highways for heavy trucks
Fee structure: Based on distance, vehicle type, and truck emissions rating
Use of revenues: 50 percent to roads, 38 percent to rail, 12 percent to waterways
Technology: GPS, GSM, and DSRC with OBU and ANPR for enforcement
Milestones: January 2005 (opened)
Managing authority: German Ministry of Transport and Toll Collect GmbH (contractor)
Other highlights: Dual automated toll collection system and manual booking system, 35 percent of trucks are foreign, less than 2 percent violations
Results: Empty trucks declined by 7 percent, 58 percent shift from dirtier to cleaner trucks, revenues of $5 billion in 2008

All trucks, irrespective of national registry, are tolled based on the number of axles, vehicle emissions rating, and distance traveled.

> The idea of a distance-based charge was conceived in 1989. Studies were conducted subsequently, and in 1995, based on the recommendations of the German High Commission (Paellman Commission) on Financing of Federal Transport Infrastructure, the federal government decided to introduce distance-based tolls. Initial opposition turned into acceptance because tolling of all heavy goods vehicles was considered fairer for German trucks vis-à-vis foreign trucks, which account for about 35 percent of all of the country's truck traffic.—Robinson, 2008

The key policy objectives were to raise revenue by imposing a national distance-based, user-pays infrastructure fee for trucks. The pricing is based on the distance traveled, vehicle emission class, and the number of axles. In addition to generating revenue, the objectives of the system were to create incentives to shift freight truck traffic to rail and waterways, promote the use of cleaner truck technologies, encourage more efficient routing and scheduling of trucks, and provide funding for maintenance and expansion of transportation infrastructure.

The system covers about 7,700 road miles (mi) (12,392 kilometers (km)), including 30-plus mi (48 km) of local roads and more than 2,460 access points (figure 18). Truckers have two payment methods (see figure 19):

- An automated system using a combination of satellite-based (GPS) tracking via onboard units (OBU), mobile communications (GSM), and DSRC for enforcement interrogation
- A manual booking system, which allows trucks without OBUs to book their travel via the Internet or through roadside terminal kiosks available to arrange for payments

Both payment methods are enforced through an ANPR system housed at 300 control point gantries, supplemented by 280 mobile enforcement vehicles. The dual system of stationary and mobile enforcement ensures that violations are kept to a minimum. Initially predicted to be 5 percent, the actual violation rate is less than 2 percent. Fines are €400 (US$580) for intentional violations and €200 (US$290) for unintentional violations. The maximum fine is €20,000 (US$29,000), with the responsibility for fines split equally between the driver and shipper.

While 90 percent of all payments are made with OBUs, the remaining 10 percent of the vehicles, most foreign, opt for manual booking. While manual bookings represent only 10 percent of all transactions, they represent more than one-third of total operating costs. There are 3,600 manual terminals in Germany and neighboring countries. Manual bookings require the acquisition of a ticket as proof of payment before using the autobahn. The system was designed to ensure nonstop, highway-speed tolling and has achieved this objective.

Figure 18. German autobahn system.

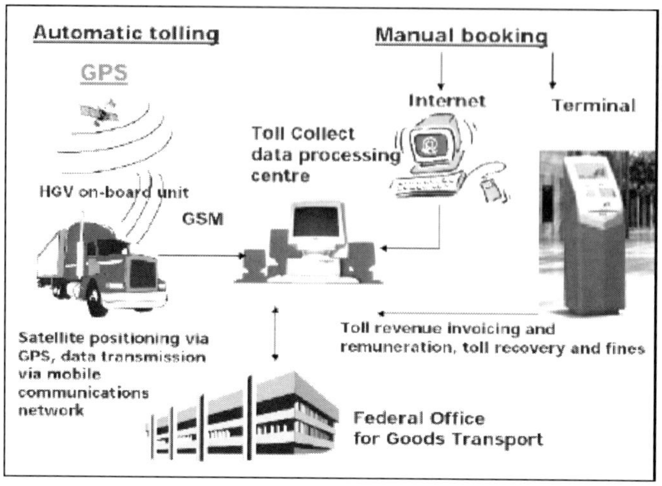

Figure 19. Automatic tolling and manual booking processes.

The OBU calculates the toll, adding segments together and charging the appropriate unit rates. As soon as the accumulated cost reaches €20 (US$29), the OBU communicates the cost information to the central processing unit. This threshold is established to avoid the costs of more frequent communication. The OBU also sends the data every 72 hours, or whenever the truck leaves the country. The OBU has three communication channels:

- GPS is used for georeferencing of when a tolled route is entered. Typically, three to four satellites at a minimum are required for accuracy.
- A DSRC module, employing infrared microwave technology, provides interoperability and communication with enforcement gantries while also supporting positioning.
- GSM communicates from the OBU to the central computing center. The GSM is a bidirectional communication stream through which the data processing center is connected to the OBU.

New tariffs are communicated to the OBU via this capability, with tariff tables stored on the OBU. The OBUs were specially designed and are provided by two competing vendors, which was a requirement of the government contract.

The OBUs will be a platform for future value-added services, but they only provide toll collection functionality today. Toll Collect is limited to toll collection functionality for OBUs by virtue of its government contract and cannot develop value-added services for the in-vehicle units. The cost of installing the OBU is paid by the hauler, and costs vary by country. It costs about €50 to €60 (US$73 to US$87) for installation. Many new trucks come equipped with the cables and connections for OBUs preinstalled. Toll Collect left the whole business of OBU installation to the private sector. The cost of the OBU, about €250 (US$363), has fallen significantly from the early days of the system's introduction.

A European Union directive lays out the principles for toll rates and revenues. Toll rates are based on the cost of the infrastructure on a life cycle basis. The toll rate is related to the cost of construction and periodic infrastructure upgrades and maintenance required. Toll rates range from about €0.141 to €0.288 per km (about US$0.327 to US$0.669 per mi). The average toll rate is €0.163 per km, or about US$0.378 per mi. Certain European Union directives restrict the highest toll rate from being more than twice the lowest toll rate. Annual gross revenues have grown steadily from €2.86 billion (US$4.1 billion) in 2005 to €3.5 billion (US$5 billion) in 2008.

To comply with European Union Directive 2004/52/EC, which requires equal treatment of foreign vehicles, accommodations were made to provide a manual payment option at the German border. While manual bookings represent about 10 percent of all transactions, they represent more than one- third of total operating costs. Those who use manual bookings have 3 days to make their trip once registered at one of the 3,600 terminals in Germany and neighboring countries. An online manual booking option is also available, but not well used. Truckers using the manual method are considered hard-core users and do not appear willing to install an OBU.

The GPS-based tolling system was procured through a public-private partnership (PPP) formed between the Ministry of Transport, Building, and Housing, responsible for contracting and system regulation; the Federal Office for Goods Transport (BAG, figure 20), responsible for enforcement; and Toll Collect GmbH, a private-sector joint venture responsible for the system's design, implementation, operations. and maintenance. Toll Collect is a private joint venture comprised of Daimler Financial Services (45 percent), Deutsche Telecom (45 percent), and Cofiroute (10 percent). Cofiroute provided the international tolling experience dictated by the German procurement. The Ministry of Transport entered a 15-year design-build-finance-operate contract with Toll Collect GmbH. As part of this procurement, Toll Collect paid for the system's capital cost and is required to maintain system accuracy and availability at or above 99 percent. The specifics of the PPP contract are not publicly available because of confidentiality agreements that protect sensitive details related to Toll Collect's proprietary software, technology, and business processes.

Figure 20. Federal Office for Goods Transport mobile enforcement vehicle.

Toll Collect received €598 million (US$868 million) in compensation for services in fiscal 2007–08. This includes operations of the automated and manual toll collection systems; service fees for payment providers; operation of the toll terminals, enforcement gantries, and mobile enforcement equipment; mobile communications; system depreciation; and net income before taxes and interest. In 2009, it was expected that 11 percent of total toll revenue would be required for Toll Collect's compensation. While it is difficult to ascertain specific operating costs and profit for Toll Collect, annual operating costs are estimated at 15 to 20 percent of revenues, about US$0.057 to US$0.076 per mi. Costs of operations have been reported in the media to be as high as 30 percent of revenues, which would be about €1.05 billion for 2008. Toll Collect officials claim this is misleading because the manual system is responsible for €140 million of the cost and depreciation is €120 million.

Revenues are allocated by Parliament using the framework of 50 percent to roads, 38 percent to rail, and 12 percent to waterways. Initially, to gain support from the German trucking industry, these additional revenues were added to the general fund allocation for roads, rail, and waterways. In addition, a freight mitigation fund was created to help truckers purchase more environmentally friendly equipment and provide training. An annual appropriation of €560 (US$815 million) derived from the net HGV toll revenues was budgeted by Parliament to support this harmonizing fund. However, the level of funding allocated from the general fund has been reduced by an amount about equal to the new revenue generated by HGV toll collections. Although the harmonizing fund continues to be fully supported, the promised funding increase for roads using revenues from HGV tolls has not materialized because of significant federal budget shortfalls. Since roadway funding continues to be flat, the original promise to dedicate all new revenues generated by HGV tolling to the transportation system has not occurred.

Initially, there were concerns that HGV tolling would divert freight trips from the autobahns to toll-free local routes. To address these concerns, three connector segments that total a little more than 30 mi (48 km) of nonautobahn roadways were integrated into the HGV tolling system. Initially, some truckers tried to bypass the HGV tolls by using local routes, but they quickly realized that the cost in time and additional operating expenses exceeded the cost of the toll. Diversion to alternative routes faded and the proportion of trucks shifting from the freeway system to local roads is now negligible.

In part because of the imposition of higher tolls on higher polluting trucks, a dramatic growth in cleaner trucks was observed. Based on the European Union's five-step emission classification scale (with Euro 5 being the cleanest vehicles and Euro 1 the dirtiest), the number of vehicle-miles using cleaner trucks (Euro 4 and 5) rose from 2 percent in 2005 to over 62 percent in 2009. Conversely, the number of vehicle-miles driven by dirtier trucks (Euro 1, 2, and 3) declined by 60 percent, from 98 percent to 38 percent of the fleet-miles traveled (figure 21). While the costs for operating the HGV tolling program are considerable, Germany has demonstrated that distance- based tolling can be successfully implemented at a national level to accomplish the following:

- Shift to a user-pays approach.
- Improve system efficiency by creating incentives to reduce empty truck trips.
- Reduce greenhouse gas emissions by promoting the use of cleaner trucks and truck technology.
- Generate substantial new funding for infrastructure.

Tracking about 1.5 million trucks annually, the system is flexible, scalable, and robust. Road segment distances are regularly audited to ensure accuracy. In addition, road segments can be added, time-of-day toll rates applied, and differentiation by road type introduced without the need for costly roadside infrastructure. While there is no public interest to expand the HGV tolling system to all vehicles, the technical feasibility of adapting the system to accommodate such an effort would be relatively straightforward.

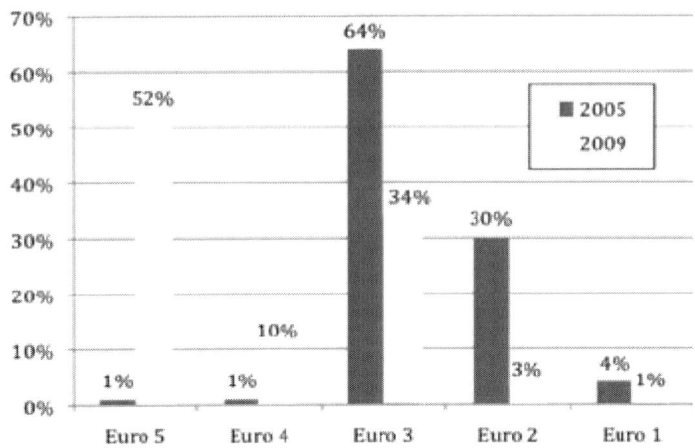

Figure 21. Truck vehicle-miles show a shift from dirtier (Euro 1, 2, and 3) to cleaner (Euro 4 and 5) trucks (2005 to 2009).

Czech Republic's Truck Tolling

Purpose: Revenue generation
Application: Truck-only distance-based pricing on selected national highways
Fee structure: Distance, vehicle classification, emissions rating
Use of revenues: Road and railway improvements
Technology: DSRC for toll collection and ANPR for enforcement
Milestones: January 2007 (all trucks 12 metric tons or more pay toll), January 2010 (expansion to include trucks 3.5 metric tons or more)
Managing authority: Czech Ministry of Transport
Other highlights: 40 percent foreign trucks, implemented in under 10 months, special law prohibiting truck operations on Sundays and peak times on Friday evening and Saturday morning
Results: Average toll rate of CZK4.05 per km (US$0.36 per mi) for highways and CZK1.90 per km (US$0.17 per mi) for first-class roads, revenues of CZK6 billion (US$340 million) in 2008

Revenue Generation: Czech Republic's Truck Tolling

Like Germany, the Czech Republic's central location in Europe draws high volumes of through truck traffic both east-west and north-south. About 40 percent of the trucks using the

Czech highway system are foreign based. Thus, the primary goal of the Czech truck tolling program is to generate revenue from foreign trucks that were perceived to not be paying their fair share of system costs, including infrastructure wear and tear. The tolling program began operations in January 2007 and uses a transponder-based system relying on DSRC-based tolling points on the highways' main line to calculate distances. ANPR equipment is employed for enforcement. At the outset, the toll applied only to trucks weighing more than 12 metric tons. In January 2010, the truck toll was applied to all commercial trucks weighing more than 3.5 metric tons.

The Czech government approved moving forward with the truck tolling program in 2004. The procurement documents for the tolling system were issued in 2005 with four suppliers responding. After the selection of the winning bid, the unsuccessful bidders filed objections and complaints.

Construction of the phase I tolling system was scheduled to begin in January 2006, but was delayed by the unresolved complaints surrounding the procurement. On March 29, 2006, the Czech Office for Economic Competition ruled in favor of the selected bidder, and a 10-year design-build-finance operate contract was signed. The procurement delays forced the stage I construction period to be reduced to 9 months. This aggressive delivery timetable was achieved by using a 5.8 gigahertz system similar to one recently deployed in Austria. Pilot operation of the electronic tolling system started in the Czech Republic on Dec. 15, 2006, and the system was launched on Jan. 1, 2007. The open road tolling system was designed and implemented based on European Union directives 1999/62/EC, 2006/38/EC, and 2004/52/EC, which address environmental consequences of roadway use, cost constraints, and interoperability of electronic toll services.

The Czech tolling system has the lowest rates of those charged in central Europe, significantly lower than the roadway charges in effect in Germany and Austria. As a consequence, there has been some commercial traffic diversion of trans-European cargo movements from other countries to the Czech highways.

Revenues from the program generated about 6 billion Czech korun (CZK) (US$340 million) in 2008, with an average toll rate of CZK4.05 per km (US$0.36 per mi) for motorways and highways and CZK1.90 per km (US$0.17 per mi) for first-class roadways (i.e., principal arterials). The Czech government requires that all trucks using Czech roads install an OBU. Certain types of vehicles, including first responders and law enforcement, are exempt from the toll, but are still required to register and install an OBU. While specific system operating costs are not publicly available, it is estimated to be about 30 percent of gross annual revenues.

Truckers or their firms submit a deposit of CZK1,550 (US$88) to the government, which pays for the cost of the OBU, its installation, and administration. In December 2009, 420,000 transponders were in use with another 100,000 transponders expected to come online in 2010 because of the change in the gross vehicle weight threshold subject to a charge from 12 metric tons to 3.5 metric tons. Changes to the toll rate are proposed by the transport minister and require approval by the Czech Parliament.

When the scan team met with Czech tolling representatives, national laws were being debated that impose unique restrictions limiting truck movements on Friday afternoon and weekend days to preserve roadway capacity for nonfreight trips on weekends. Resistance to these measures was significant from freight interests unable to work a full week. Other problems with the Friday restrictions were lack of capacity for truck parking and incompatibility with driver restrictions in other countries. The restrictions on Friday truck traffic reduced traffic congestion on Czech highways 50 to 60 percent, according to Czech officials. Toll rates were changed on Feb. 1, 2010, to increase regular truck toll rates by 25 to 50 percent on Fridays from 3 to 9 p.m. in an apparent effort to use market pricing rather than vehicle restrictions to manage commercial traffic demand during that peak congestion period.

This implementation of truck tolling on freeways and expressways is considered the first phase of a more comprehensive road pricing system (figure 22). Future plans (phase 2) are to extend tolling to 800 km of additional roadways by 2017 by including some arterial and local roads. Capital costs for the initial implementation (phase 1) and future expansion (phase 2) of the system are estimated at about CZK18 billion (US$1 billion). When complete, phases 1 and 2 will include 1,120 mi (1,802.4 km) of expressways and freeways and 43 mi (69.2 km) of connecting main roads. This expansion of the existing system will seek to employ microwave technology and GPS-based OBUs because the cost associated with installing additional gantries on arterials and local roads is considered excessive.

Demand Management and Revenue Generation: The Netherlands' Proposed Distance-Based Tax

The Netherlands is a small, densely populated country with an integrated, multimodal transportation system, but it nonetheless experiences highly congested roads. As a consequence, the Netherlands has been planning an ambitious national road pricing program. The plans have been to introduce a countrywide distance-based road user fee for trucks by 2012 and to expand the system incrementally to all vehicles (about 8 million) by 2018.

Figure 22. Czech truck tolling network.

In late 2007, the Dutch Cabinet decided to implement a national road payment based on the following policy goals:

- Improve mobility and accessibility to benefit the economy.

- Develop a more fair system that focuses on use of the road system rather than on vehicle ownership and replace current license fees and vehicle taxes.
- Enhance the environment.
- Improve road safety.

The Dutch Cabinet considers a distance-based road-user fee to be a more transparent and equitable method to fund the transportation system and effectively manage congestion. The new system is designed to be revenue neutral by phasing out high license fees and other vehicle ownership taxes and introducing a per-kilometer charge in their place. While the precise price per kilometer is still in planning, the expected tariff for autos at the start is €0.030 per km (US$0.07 per mi), with the rate expected to reach €0.067 per km (US$0.155 per mi) at full implementation. The lower rates at the program's start reflect the fact that vehicle ownership taxes will be phased out over time and user fees will increase. The per-kilometer tariffs for other vehicles at implementation are planned to be €0.017 per km (US$0.039 per mi) for commercial vans, €0.028 per km (US$0.065 per mi) for buses, and €0.024 per km (US$0.056 per mi) for trucks. The relatively low rate for trucks corresponds to the fact that trucks pay relatively lower rates for the vehicle taxes that will be replaced by the road charging system. In addition to pricing by vehicle type, the system was planned to take into account vehicle emission class and time-of-day pricing for peak and off-peak travel. All revenues will be dedicated to the road system to offset existing license and registration fees.

The Netherlands' Kilometer Tax

Purpose: Improve mobility, manage demand, and shift to user-pays approach (primary)
Application: National distance-based pricing of all roads for all vehicles
Fee structure: Based on distance and vehicle type with option to include emissions class and time of day
Use of revenues: Replacement of vehicle ownership taxes now in place
Technology: GPS, ANPR, GSM
Milestones: 2012 launch (trucks only), 2018 (all vehicles)
Managing authority: Dutch Ministry of Transport
Highlights: In planning stages and under development, unique requirements that minimize system installation and operating costs, revenue neutral (offset by reduction in other transportation taxes)
Forecasted results (2020): 10 to 15 percent reduction in vehicle-miles traveled, 10 percent reduction in CO_2 emissions, 6 percent increase in public transit use

Based on 2020 forecasts, the distance-based tax would result in the following:

- Ten to 15 percent reduction in kilometers traveled
- Travel time savings of 40 to 60 percent
- Ten percent reduction in CO_2 and PM_{10} emissions by passenger cars
- Six percent increase in miles using public transport
- Seven percent increase in traffic safety

The method for toll collection and vendor selection has not been decided, but the Dutch high-level system design envisions a satellite-based (GPS) technology combined with in-vehicle equipment and mobile communications. The procurement approach will be to employ open systems design standards that comply with the European Union's future standards for tolling interoperability. The intent is to enable free marketplace access to the road-pricing system by certified suppliers of products and services. The procurement will involve market-based parties whenever possible, leaving detailed system designs, system implementation, and equipment installation in vehicles to private sector companies and giving the government the role of ensuring installation quality. This open approach to design, implementation, and installation is intended to create market competition and provide private interests with the incentive to develop value- added services for the equipment being deployed. Similar to what is occuring in the mobile device industry, this would enable a wide variety of developers to provide enhanced applications and services that deliver the basic toll collection function, but also provide greater value to users. This approach is expected to yield revenues that may offset the cost of initial implementation and ongoing operating costs.

Administrative costs are a major concern for the Dutch. To manage costs, they have looked aggressively at technologies and business processes to minimize them. The Dutch Parliament has set a goal for operating costs to not exceed 5 percent of gross revenues and for OBUs to cost less than €100 (US$145) installed. At the proposed fee structure, it is estimated to cost US$0.0075 per mi, based on 5 percent of US$0.155 per mi, to collect and enforce the Dutch kilometer tax program.

Figure 23. Congestion on the roadway system in the Netherlands.

The Dutch have also created a €100 million (US$145 million) fund to support innovative congestion management projects from 2007 to 2011. This fund is intended to provide more immediate, short-term congestion relief through road pricing pilots in selected corridors without adding new roadway capacity. Next steps include extensive consultation, amendment of legislation, and development of a more detailed migration plan from vehicle taxes to the distance-based tax. While the Dutch political climate is unsettled because of upcoming elections, steps have been taken to conduct a large-scale system test in 2010.

On Feb. 20, 2010, about 2 months after the scan team returned to the United States, the three-party majority coalition comprised of the Christian Democratic Alliance, Christian Union, and Labor Party collapsed over the issue of continued Dutch troop support of North Atlantic Treaty Organization efforts in Afghanistan. The coalition government held only 46 of the 150 Parliament seats with the departure of the Labor Party and was relegated to a caretaker government role until the June 2010 elections. With Camiel Eurlings' resignation in March 2010 from the post of transport minister, the fate of a future large-scale test and overall implementation of the Netherlands' distance based tax was uncertain.

3. Major Findings

The countries the scan team visited are ahead of the United States in broadscale road pricing implementation. They each provide valuable case studies for learning for U.S. transportation professionals and decision makers. Overall, the experience in each host country proved that road pricing is an effective tool to manage demand and raise revenue.

The scan team developed the following nine major findings:

1. Host countries and regions with clearly defined and well understood policy goals were able to achieve their targeted outcomes most effectively.

The city-center urban road pricing programs visited all targeted congestion mitigation as a central goal. Generally, road pricing was one element of a larger program of initiatives working collectively to address traffic congestion and its impacts. Programs that sustain a focus on traffic congestion have sustained traffic-reduction benefits.

- Singapore identified clear transportation goals as a critical foundation for urban development and economic growth plans and has maintained this focus for many years. Congestion management objectives include road pricing tied directly to targeted minimum- speed thresholds for urban streets and arterials. In addition to the ERP system, congestion management is also addressed through multimodal transportation investments as alternatives to driving, parking management systems to facilitate identification of available parking, and national quotas that cap increases in vehicle ownership. Highly integrated land use and transportation planning also support congestion management objectives.
- Stockholm's congestion tax was designed to reduce congestion in the city center in a manner that promoted public acceptability and fairness through reasonable toll prices, daily maximum charges, expanded transit and park-and-ride services as alternatives to driving, and exemptions for Lidingö Island traffic unable to avoid traversing the city center. The congestion pricing program improves the quality of life in the city center for residents and travel options for drivers. In Germany and the Czech Republic, the distance-based charging of commercial vehicles was designed to capture revenues aligned with the infrastructure life cycle costs required to maintain roadways for HGV use. Political decisions in Germany to divert existing general

funds from transportation to address other budget shortfalls contradicted commitments to use road-pricing revenues to augment existing infrastructure investment, fueling resentment and skepticism from trucking interests.

- The Germans and Czechs viewed their pricing programs as an opportunity to capture road revenues from foreign through truck traffic, but found that manual booking systems for those without in-vehicle units are expensive to operate.
- The Netherlands plans to introduce a distance-based charge for all vehicles and roadways. This kilometer- priced system would replace motor vehicle taxes on passenger cars and commercial trucks and eliminate vehicle purchase taxes on passenger cars and motorcycles. As such, roadway network users would pay for vehicle use, not vehicle ownership. Existing fuel taxes would remain in place. Thus, the government's policy goal is to improve mobility and quality of the environment. The Dutch government believes a true user tax would be more effective for managing demand than the current ownership taxes, which have the effect of overcharging car owners who do not drive much. For example, owners of classic cars pay an ownership tax even if they do not drive the cars. The universal user-pays principle in the Netherlands is expected to generate strong environmental benefits from trip consolidation and alternative mode choices. The environmental benefits achieved by the Germans and Czechs are evident through price incentives for cleaner vehicles and a reduction in empty truck movements.

2. A large-scale demonstration project is a powerful tool for building public acceptance, allowing people to experience the benefits of congestion pricing.

Stockholm's trial of the congestion tax system from January to July 2006 demonstrated the benefits of congestion pricing firsthand. Sequencing a referendum vote after the trial concluded was instrumental in garnering public support. The pilot demonstration also provided technical and administrative staff with opportunities to refine the system and its performance, streamline business processes, and reduce operating cost.

3. Thorough planning and performance measurement pay benefits in ensuring achievement of overall goals, managing the pricing program as an element of overall transportation system performance, and directing implementation and operations effectively.

Comprehensive network planning was integral to the preimplementation efforts for the road pricing systems examined on this scan. The best-in-class road pricing programs have integrated public transport options into their planning and preimplementation actions.

- London, Singapore, and Stockholm made significant advance investments in transit equipment, facilities, and services.
- In planning the Stockholm system, internationally recognized traffic experts were retained to measure network effects of various configurations of the charging zone to ensure that there were no unintended effects outside the congestion charging zone.

- The Netherlands has undertaken comprehensive planning exercises to look at network effects of proposed tariffs across several modes, as well as the operating performance of the network when travel demand is redistributed by time of day.
- Singapore uses advanced analytics and traffic models to better understand the network impacts of pricing on parking and transit.

Performance measurement is key to managing and maintaining goal attainment.

- All new pricing systems adopted direct performance measurements of traffic reductions, travel speed increases, mode shift, and clean vehicle adoption, as well as estimates of business impacts and emissions reductions.
- Singapore's ongoing management of its congestion charge includes quarterly verification of travel speeds and refinement of prices to ensure that 85th percentile travel speed standards are maintained on two different classes of roadways.
- The Netherlands has adopted comprehensive risk analyses to manage program schedules and budgets.

Postimplementation planning and performance assessments focused on the right measures have ensured cost efficiency and operating effectiveness.

- Stockholm has benefited from ongoing assessments of system redundancies and business practices to reduce operating costs while maintaining system performance. Changes to payment methods, payment processes, and image processing have saved systems and operating costs.
- The Czech Republic is seeking relief from contractually high DSRC OBU costs by planning system expansion to employ GPS and microwave technologies.
- Germany's truck tolling system suffers from high costs for manual bookings, but the joint venture contract incentivizes system accuracy, not cost performance.

4. Linking the pricing structure to the benefits the user receives contributes to public acceptance and helps avoid potential negative impacts from traffic diversion.

To maintain support for road pricing, some of the sites studied attempt to connect the pricing structure to the benefits received by the toll payer.

- In Sweden, toll rates on new roadway infrastructure are set at levels that reflect the value of the reduced travel time and operating costs compared to existing non- priced routes. While this strategy may not fully recover the costs for the new infrastructure (thus requiring public subsidy), it ensures that the new infrastructure will be used optimally (i.e., traffic will not stay on existing free routes simply to avoid the toll). In Stockholm, the price schedule for the cordon charges was set to reflect the expected benefits to those who would pay the new charges. The concern was that if the public perceived tolls to be too high, they would not accept them. Toll rates were set at the minimum levels needed to manage congestion, not to achieve a specific revenue target. The congestion tax rates were set to match the value of time saved. Since its introduction, the congestion charging rates have not been changed. While there is a recognition that future increases will be needed to address growing travel demand and inflation, public acceptability of any rate changes is an important consideration. Because the congestion charge is defined as a national tax, it can be changed only by an act of Parliament, which makes it less nimble.

- In Singapore, the price is reviewed and can be adjusted every 3 months to ensure that targeted optimal speeds can be maintained for at least 85 percent of vehicles that pay the toll. The concern is that the public obtain value in the form free-flowing traffic in return for the tolls paid. In the early days of operation, LTA attempted to set toll rates to achieve the targeted speeds on average. It quickly learned that this meant that only about half of the toll-payers received the targeted service levels, and the public perceived it had not received value for tolls paid. This led the authority to institute the 85th percentile standard. All net funds collected via ERP are returned to the general fund and redistributed to road users in the form of vehicle ownership tax rebates, which further emphasizes that the purpose of road pricing is not to generate revenue, but to improve service levels during peak hours. Singapore's ERP has the most dynamic and flexible pricing structure of the sites visited.
- The German truck toll rates for use of the freeway system (i.e., autobahns) are low enough that little or no truck traffic is diverted to toll-free alternatives. While there was some diversion immediately after implementation in 2005, truckers quickly realized that time and operating cost savings on the autobahn system more than compensated for new toll charges for using the autobahns. The Germans set the truck toll rates to average €0.163 per km (US$0.378 per mi) to capture the impact of HGVs on the transportation system. The fairness of the toll rates as charges for services is reinforced by the way toll rates are calculated. To determine the average toll rate, the estimated infrastructure cost imposed by HGVs (estimated in 2010 at €5.2 billion) was divided by the total number of HGV km on the highway system (29.8 billion km). This is calculated at €0.174 per km, so the €0.163 per km actual average toll rate is perceived to be fair. It captures more than 90 percent of the estimated HGV impacts on the highway network. In addition, the €560 million (US$815 million) annual harmonization fund for truckers is dedicated to trucking community use for safety training and equipment purchases, which provides a direct nexus between the funding source and the use of funds.

5. public outreach and communications were key components of the program at every stage: before the implementation decision, during the program design process, and during the operational phase.

Both London and Stockholm had years of public debate about congestion charging before the political decision to implement was made. London's program benefited from promotion by business groups concerned about congestion, while the Stockholm program was spearheaded by environmental groups. Both programs were designed to address public concerns and include a number of exemptions and discounts to mitigate negative impacts on particular segments of the public.

After several attempts to implement a distance charge, the Dutch realized that proactive stakeholder outreach during the planning and concept development stage is essential. Over the past 2 years, staff and leadership at the Dutch Ministry of Transport have invested heavily in public outreach and education. By engaging in a thorough and thoughtful planning and public involvement process, the Dutch developed clear, salient, and timely messages about the purpose and benefits of pricing. A key message is "drive less, pay less."

In Singapore, key messages continue to be conveyed to the public to ensure continued support. These messages include the following:

- "Keep roads free-flowing"
- "People-centered transportation"
- "Public transit is a viable choice"

As system expansion continues, before a new gantry is installed, there is extensive outreach in affected neighborhoods to address concerns. For example, the times for operation of expressway pricing were modified to respond to public comments.

The Czech and German programs achieved support from local truckers by emphasizing the effect on leveling the playing field with foreign haulers and the fairness of the user-pays concept.

6. Open-source system designs offer long-term advantages in leveraging market competition to manage costs of implementation and operations, ensure system flexibility and scalability, and establish a foundation for system interoperability.

The Dutch plan will establish standards and requirements that will allow multiple vendor solutions to create a competitive environment.

- The Dutch procurement will encourage market engagement in all aspects of the system, driving down costs for system implementation, equipment, and in-vehicle installations.
- Through open standards and private sector engagement, the Netherlands will encourage private value-added services for onboard devices that help defray operating costs as well as drive consumer adoption and public acceptance of road pricing.
- The Dutch goal is for system operating costs to not exceed 5 percent of gross revenues.

Singapore's second-generation smart card for the in-vehicle unit financial purse is designed to be interoperable with the transit fare media and parking payment systems, as well as accepted for retail purchases and linked to bank accounts.

7. Interoperability among states and countries is recognized as a critical issue that needs to be addressed at high levels.

The European Union has adopted Directive 2004/52/EC, which outlines requirements for member countries to adopt interoperable standards (i.e., European Electronic Toll Service (EETS)) for electronic tolling, allowing one vehicle to pay road user fees anywhere in the European Union via one contract and with one OBU.

- Technical, administrative, and legal hurdles have made advancing interoperability time-consuming and challenging. The European Parliament and European Union Council approved the directive in 2004. Five years later, a decision on the EETS definition was approved in October 2009.
- Existing systems with large sunk costs in proprietary applications and equipment heighten the challenge of transition.

Interoperability addresses technical, business, administrative, financial, and legal issues, requiring thorough treatment and multidisciplinary expertise.

Intergovernmental coordination in sharing national vehicle registry information between agencies is key to today's operations and enforcement and for interoperable systems of the future.

- All sites visited have procedures in place between agencies in their own country to share vehicle registry data for easy applications of license plate imaging for invoicing and violation processing.
- The use of ANPR technologies for enforcement is ubiquitous. It is used in the Czech Republic, Germany, London, Singapore, and Stockholm, but more agreements to share information across borders are needed.
- Clearly defined and well-understood policy goals in Stockholm guided decisions large and small, such as revenue use, rate schedule, and appropriate technology solutions.
- The Netherlands plans to use a tax-consolidation approach to transition fragmented vehicle ownership taxes to a single distance-based user fee.

8. Equity and privacy concerns are addressed by host countries through exemptions, revenue use, technology, and business rules.

- Exemptions are used in London and Stockholm to help address issues of equity. In addition, the emphasis on using toll revenues to fund transit sends a strong, clear message about equity and the project purpose.
- Privacy was elegantly handled by Singapore's use of a smart cash card that does not contain user data. The primary data on the smart card is the account balance, which can be used to pay for parking or other amenities.

9. The urban area pricing projects integrated public transit investments and land use planning to manage congestion.

- Stockholm and London made robust investments in public transit and alternative modes leading up to and following the introduction of road pricing. In Singapore, officials adopted and committed funding to realize a master transportation plan that integrates road pricing, transit, roadway expansion, and land use.
- The coordination of road pricing policy with public transportation investments is best accomplished by a single entity. In London, TfL is responsible for implementing the mayor's Transport Strategy for London and managing transportation services for all

modes of transportation and throughout the city. In Singapore, LTA plans the long-term transportation needs of Singapore for those who drive and those who take public transportation. The Swedish government is in the process of consolidating its transportation agencies to bring all modes under one umbrella.

4. Lessons Learned

In addition to the major findings cited in the last chapter, the scan team found it useful to consider a number of lessons learned that are characterized by functional categories and business disciplines. The seven categories of lessons learned discussed in this chapter include the following:

1. Political and policy considerations
2. Legal and institutional issues
3. Planning and performance measurement
4. Procurement
5. Technology
6. Operations and enforcement
7. Outreach and communications

The lessons learned are intended to provide more in-depth discussions of the findings from discussion with the host countries. They are organized to enable subject matter experts to examine the areas of greatest interest most effectively.

1. Political and Policy Considerations

A. Strong champions were critical to initial implementation of road pricing in several countries. the successful programs blend government leadership to guide executive and legislative government agendas with strong program manager champions in the implementing agency.

Executive champions, such as former London Mayor Ken Livingstone and Stockholm Mayor Annika Billstrom, led the charge to implement road pricing to improve livability and sustainability in their capital cities, despite fervent opposition and divided public support. Their vision, understanding, and political aptitude created the executive mandate for road pricing and laid the framework for the political coalitions and legislative changes required to institute congestion pricing. In Stockholm, a strong city manager directed the congestion pricing program implementation, while the London program management benefited from key leadership appointments at TfL. In both cases, agency leaders worked in concert with elected city executives to ensure implementation aligned with overall objectives of the pricing program.

After the scan team's visit to the Netherlands, the ruling government coalition dissolved and the Dutch champion of road pricing, Transport Minister Camiel Eurlings, resigned.

Eurlings set road pricing as a national priority, so his departure created a vacuum in leadership, project guidance, and legislative action in early 2010. A number of political parties still support the distance-based road pricing system, and the leading party in the former government coalition, the Christian Democratic Alliance, has said it will recommend a revised version of the distance-based charge in an upcoming proposal. Nonetheless, the Netherlands' comprehensive, multidisciplinary approach to guiding implementation has suffered a setback because of changes in leadership and champions.

In contrast, the consistency of Singapore's government leadership and singleness of purpose at LTA to strive for a sustainable and effective transportation system have been hallmarks of the road pricing program's success.

B. Simplicity matters on policy goals, messages, business rules, and technology solutions.

Stockholm's Lidingö Rule was created to provide an exemption from the congestion tax to Lidingö Island residents, who must cross the Stockholm charging zone to access the national road network. The rule allows free passage for any vehicle that exits the charging zone within 30 minutes of entering the zone at the Lidingö Island control point or vice versa. This exemption maintained a policy of free access to the national highway network, but added more than US$20 million in capital and operating costs. It also added considerable system complexity to toll operations because every vehicle needs to be tracked to verify if it is making a qualifying trip for this exemption. If one gantry fails to operate properly, all vehicles during that period must be assumed to qualify for the Lidingö Rule and are exempt from the congestion charge.

Stockholm advanced simplicity in its permanent system implementation through a number of revisions during the 11 months between the referendum and the activation of the permanent system. The first was a conversion to fully video-based transactions, eliminating the cost of transponders, DSRC field equipment, and operations to support a second payment method. Second, a rule requiring that congestion taxes be paid by the day after the charge was incurred was revoked. This allowed the elimination of a costly system of payment channels, including Internet, retail locations, and stand-alone kiosks. Instead, drivers receive a monthly billing statement from the state for their congestion tax transactions, which saves significant operating costs and simplifies the system for drivers. Finally, these changes were accommodated by an alignment that placed the congestion tax program and the national vehicle registry under the same organizational oversight, creating efficiencies in data sharing and invoicing.

To comply with European Union Directive 2004/52/EC, which requires equal treatment of foreign vehicles in domestic road pricing programs, the German truck tolling system was required to establish a manual payment option for trucks that did not acquire an in-vehicle unit. While only 10 percent of truck transactions are paid via the manual system, manual transaction processing represents over one-third of the total operational costs.

German business objectives for the truck tolling system included the ability to have a flexible road network for expansion and scalability, as well as a flexible tariff schedule. These objectives were achieved through a complex remote device management process by which changes in the road network and pricing are centrally propagated to all users' OBUs in a seamless and consistent manner. The system's sophistication requires maintenance of a complex network of geographic data in five distinct layers to detail roadway links, distances, locations, and related tariffs and decision analytics. System maintenance requires regular updates of 640,000 OBUs with 600 geographic data changes per year and 400 annual tariff changes. While the system meets the goals established, it requires a high level of technical input and system administration to ensure accuracy of OBU functionality in a secure manner.

London began its congestion charging program with a simple set of policy objectives that prioritized reductions in traffic volumes and related congestion. The program's well-documented success in both the original charging zone and the western extension created roadway capacity. This newfound capacity provided an opportunity to address a larger set of urban needs, including new bicycle and pedestrian uses of streets in the city center and infrastructure modernization such as water supply. These actions have consumed street capacity to the extent that traffic congestion in central London is reportedly back to levels experienced before congestion charging. By attempting to address a broader set of objectives after successes in congestion reduction, the London system has lost ground in its simple origin of congestion reduction.

C. Equity concerns and perceptions of fairness can be addressed on multiple levels.

Equity concerns are typically addressed via exemptions and discounts to the congestion charge. For example, traffic passing through Stockholm's city center is exempt from the congestion tax to ensure that residents of Lidingö Island have access to the national highway network. Among its various discounts, London also maintains a 90 percent discount for residents in the central London congestion charging zone.

In London, Singapore, and Stockholm, where demand management was a prime objective for urban road pricing, significant investments in transit were made in tandem with the implementation of the road pricing program to ensure viable options to driving. Furthermore, the net revenues, directly or indirectly, are used to fund future multimodal enhancements.

The use of revenue has been another means of addressing equity and perceptions of fairness. In Germany, a truck harmonization fund created from road pricing revenue pays for new truck equipment and training for cargo haulers. In Singapore, net revenues not invested in transportation projects are returned to motorists through rebates in vehicle taxes.

D. Political timetables and deadlines create opportunities and challenges for implementation.

The Stockholm demonstration period was a prescribed timeframe in which the area-wide pricing system needed to be operational on a trial basis before the planned referendum. While the schedule created urgency for system delivery, it also caused problems because of legal challenges to the procurement. The Stockholm trial was delayed 5 months from its original planned startup, requiring extraordinary effort by business consultants, the system integrator, and agency staff to deliver the system as a 7-month trial. The Czech Republic also experienced a compressed delivery schedule because of procurement challenges. The system integrator and construction contractors had only 9 months to install and test the Czech truck tolling system.

These examples illustrate that expedited delivery of high-functioning and reliable pricing systems is possible. However, both Stockholm and the Czech Republic are managing systems that are expensive to operate and maintain. The Stockholm program is undergoing a multiyear program of business process and systems reenginee ring to reduce operating costs.

2. Legal and Institutional Issues

A. The statutory and legal authority for road pricing was an essential foundation for all programs and often requires time to establish.

Road user fees first became possible in Sweden in 1988 through a change in law, which allowed tolls on the bridges to Norway and Denmark. Stockholm's congestion-pricing proposal became a matter of national debate and legislation because the city has no power or authority to charge nonresidents.

B. The statutory and legal basis for road pricing programs has implications for system design, operations, and enforcement.

Stockholm defines its road pricing as a national tax, since the city of Stockholm has no power to tax nonresidents. As a consequence, changes to the fee schedule require parliamentary action. In London and Singapore, the congestion charge is considered a fee, which allows the rates to be adjusted by the managing authority and fines to be resolved through administrative actions. Stockholm officials report that the benefit of treating congestion charges as a tax is higher collections because taxes are paid with higher frequency than other fees.

C. Interagency agreements and relationships were evident for effective operations and enforcement.

All pricing programs reviewed on the scan have procedures and relationships in place to share vehicle registry data for operations and enforcement among agencies. Some agencies, such as the new Swedish Transport Agency, operate the priced facility and manage vehicle registration, further streamlining data sharing, billing processes, and associated costs.

D. Legal requirements governing the use of road pricing revenues can protect revenues for targeted transportation investment, but flexible provisions for revenue use have leveraged revenues to their best advantage.

In Germany, the revenue generated from the truck tolling program is legally required to be allocated to projects for roads (50 percent), rail (38 percent), and waterways (12 percent). This preserves the majority of the revenue benefit for road users, but helps meet a national goal of multimodal transportation solutions to address mobility and future capacity needs.

In Sweden, the national government has established a long-term 2010–2030 investment plan for infrastructure investment, with the equivalent of US$14.5 billion programmed from a variety of revenue sources, including the Stockholm congestion tax and plans for congestion taxing in Göteborg. The investment plan allocates 53 percent of the funding to road projects and 47 percent to rail projects. The Swedish policy of "advancement" is incorporated in the

Road Act to allow municipalities to augment national funding, making more investment possible and creating higher design and operating standards.

E. Limits to the price-setting mechanism through laws and directives may be required to gain public acceptance, but may also constrain the ability to use price signals most effectively.

A European Union directive dictates that the differential between the highest and lowest toll rate cannot be more than two times the lower value. Some European officials claim that this differential is not significant enough to manage traffic effectively in urban areas. The argument is that the limited price differential is insufficient to allow efficiencies to be realized through substantial temporal or mode shifts or avoided road trips.

European Union Directives 1999/62/EC and 2006/38/ EC cover vehicle taxes, tolls, and user charges imposed on vehicles intended for the carriage of goods by road and having a maximum permissible gross laden weight of not less than 12 metric tons. The directives establish total revenue thresholds that must not exceed infrastructure costs, which some officials view as limiting for network investment purposes.

3. Planning and Performance Measurement

A. The best planning practices address education, outreach, and stakeholder management comprehensively.

The Dutch have studied distance-based charging and engaged key stakeholders on the subject several times since the 1980s. Recent implementation planning in the Netherlands involved extensive outreach during the system design and development process with a wide range of public agencies, private interests, and industry user groups. The Dutch provided significant funding for implementation research in a US$150 million congestion mitigation program.

Singapore, on the other hand, used grassroots representatives to gauge public sentiment before expanding the charge to the Orchard Area shopping district and expressways. Consultation typically starts with core stakeholders and later reaches out to the public through communications programs.

B. For many of the pricing programs reviewed, comprehensive network planning and performance measures were integral to preimplementation efforts, as well as ongoing system management.

Most sites use advanced analytics and traffic models to better understand the network impacts of pricing on parking, transit, and system diversion issues. In planning the Stockholm system, internationally recognized traffic experts were retained to measure network effects of various configurations of the charging zone to ensure that there were no unintended effects outside the congestion-charging zone. Similarly, the Netherlands has also undertaken comprehensive planning exercises to look at network effects across several modes.

Singapore's ongoing management of its congestion charge includes quarterly verification of travel speeds and refinement of prices to ensure that 85th percentile travel speed standards are maintained on two different classes of roadways.

C. The best-in-class road pricing programs have transport options integrated into their planning and preimplementation actions.

London, Singapore, and Stockholm made significant investments in transit to ensure that those impacted by the new road-user fees would have alternatives to driving. Such plans provided adequate capacity and service levels to ensure balanced transportation network demand, and limited minimal impacts to mobility and businesses with urban charging zones.

When provisions of modal alternatives to driving are not feasible, several pricing programs have employed exemptions or discounts to the road-user charge. Stockholm exempted Lidingö Island residents who pass through the city center to access the national highway network from the congestion tax. Similarly, residents in the central London and western extension charging zones enjoy a 90 percent discount on the congestion charge.

D. Geography has played a role in the design and business rules of many of the pricing programs visited.

Stockholm's city center is an island with well-defined access points that served to define roadside equipment locations and customer understanding of the limits of the charging zone.

Both Germany and the Czech Republic are central to European goods movement, handling high volumes of out-of-country movements on their national highways. Pricing system designs and business rules were established specifically to meet objectives for both native and foreign truckers.

Singapore is an island city-state with limited land available for development and economic expansion. As a consequence, Singapore has instituted regulations and planning processes to target specific land uses, encourage high-density development linked to transit, and manage the demand for new vehicles through its vehicle quota service.

A public consultation process preceded the London congestion charge, leading to many exemptions from the charge. A public consultation held after Mayor Boris Johnson took office recommended discontinuing the congestion charging in London's western extension.

4. Procurement

A. Open-source system designs provide significant advantages.

Open-source system designs create a competitive bidding environment for capital and operating costs associated with initial implementation, including system development and integration, roadside and in-vehicle equipment, construction, and back office (figure 24).

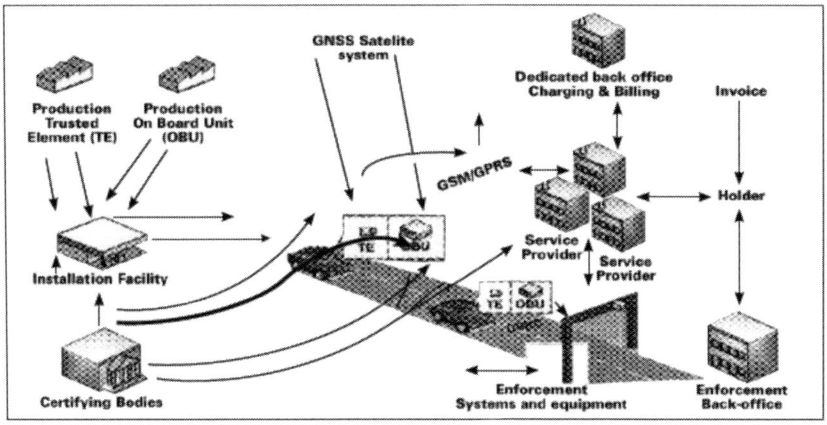

Figure 24. Example of open- source system design.

The German truck tolling system contract required a minimum of two vendors capable of providing in-vehicle units to ensure a competitive pricing environment. The Dutch system planning has expanded on this concept by proposing to allow multiple vendor solutions, creating a competitive environment for equipment and systems development and an open environment for value-added services that may defray costs and support public acceptance.

In Singapore, second-generation stored-value smart cards for in-vehicle units are designed with an open standard to be interoperable with the transit system and permit interfaces with financial institutions and businesses. The new smart cards will allow value-added services and cashless payment options for many goods and services.

B. Road pricing system contracts have been competitive and lucrative for the businesses supporting them, inviting contentious legal challenges to procurement decisions.

Legal procurement challenges from vendors in the Czech Republic, Germany, and Stockholm affected implementation schedules and some system requirements. Procurements of this type are often delayed by challenges to the selection process, so it is advisable to build in schedule contingencies to accommodate these circumstances.

The complexities of the Dutch procurement plans for open systems designs and significant private sector participation led the Netherlands to employ extensive risk assessment and cost estimation planning to assess private sector procurement options.

C. Rapid implementation schedules for pricing projects are possible, but have driven up capital and ongoing operating costs. clarity and specificity of the project scope greatly influence cost and schedule.

Stockholm worked with an open scope, which was deemed necessary to deal with the compressed delivery schedule for the trial program and evolving legislation that was not resolved at the time of the procurement. The consequence was schedule delays in the trial startup because of scope changes to access legislative outcomes and high implementation and operating costs.

The Czech Republic selected an off-the-shelf DSRC-based system that had relatively high OBU costs and a design-build-finance-operate PPP contract over 10 years. The contractor started receiving payments after 6 months' worth of revenues were generated. The German private consortium also fronted the capital cost for a 15-year design-build-finance-operate PPP contract.

Stockholm and London used more traditional design-build-operate procurement methods and found that initial capital and operating costs were high, requiring subsequent actions to reduce ongoing costs.

D. PPPs have been used in some road pricing deployments, leveraging no upfront costs for implementation, but incurring high operating costs.

The German Toll Collect consortium put up all the money required to develop and implement the German system through 2005, with no public funding. The private consortium is bound by contractual standards for availability and accuracy and is audited by the federal government on a regular basis. Toll Collect's compensation is about 11 percent of total toll revenue and includes operation of the automated and manual systems; a service fee for payment providers; operation of the toll terminals, enforcement gantries, and mobile equipment; mobile communications; system depreciation; and net income before taxes and interest. The nature of the contract provides little incentive for cost efficiency, which drives disproportionate spending to ensure performance standards.

Similarly, the Czech system is a PPP with a long- term contract that locks in relatively high costs of operations.

5. Technology

A. Business and functional requirements should guide technology selection.

Several pricing programs realized benefits from a requirements-based approach to program development and systems design. Countries that had business, functional, and technical requirements at the core of their procurement processes tended to be more successful in engaging public and private partners to achieve objectives.

The Germans chose GPS-based technologies to meet their business requirements of an easily expandable and scalable priced roadway network. The functional requirement for OBUs to be managed remotely with downloadable roadway location networks and tariff schedules drove the technology selection.

The Czech Republic opted to prioritize the implementation schedule as a critical consideration. As a consequence, the Czechs chose an off-the-shelf DSRC/ radio-frequency identification system, which was rapidly deployable. Relatively high unit costs and a long-term contract constraint have given the Czechs a reason to explore alternative technologies as they plan to expand the pricing system to new roadways.

Addressing privacy protection and the perception of privacy was a key consideration in Singapore and in the plans for the Netherlands. In Singapore, privacy concerns were mitigated by a requirement for OBUs to accept a stored-value smart card as the payment mechanism. Since the prepaid smart card is portable and does not hold any personal data, an individual's privacy is protected. Similarly, the Dutch plan to require a trusted element as a feature for privacy protection.

B. Initial technology applications often evolve after implementation, especially with the experience of full-scale operations.

Stockholm migrated from a system that employed both transponders and a video toll collection system to one that relies solely on ANPR. The dual payment methods were costly to operate and Swedish tax law required that the Stockholm congestion tax system capture photos of all license plates. After the demonstration period, the system was permanently reopened and the transponders were phased out, allowing the conversion to a monthly billing statement rather than requiring users to pay per trip.

Singapore migrated from a paper permit to a transponder-based system and is now exploring GPS technologies.

London is reviewing its charging techniques to find ways to reduce the operational costs of the system.

C. Requirements that add complexity to the collection system also have significant impacts on costs.

A fundamental decision in Singapore was to require an OBU in every vehicle, which greatly simplified technology requirements and keeps ongoing operating costs low.

While only 10 percent of transactions in the German system are manual, they account for over one-third of the total operating costs.

The Lidingö Rule in Stockholm added substantively to the systems development and processing requirements to handle the complexity of the decision algorithms required to check for exempted through-traffic movements. The increased complexity has added significantly to capital and ongoing operating costs.

D. Interoperability of road pricing systems among European Union member states is a challenge.

The European Union has adopted Directive 2004/52/ EC, which outlines requirements for member countries to adopt interoperable standards (i.e., EETS) for electronic tolling that allow a vehicle to pay road-user fees anywhere in the European Union via one contract and with one OBU. Technical, administrative, and legal hurdles have made advancing interoperability time- consuming and challenging. The European Parliament and European Union Council approved the directive in 2004. Five years later, a decision on the EETS definition was approved in October 2009.

While interoperability may be viewed as a technological concern, it interfaces with a full range of business, administrative, financial, and legal issues. Establishing standards is a critical first step, but ultimately interoperability will require multidisciplinary approaches. Existing systems with large sunk costs in proprietary applications and equipment heighten the challenge of transition.

6. Operations and Enforcement

A. Enforcement is key to ensuring a financially viable and fair system.

Video enforcement is an essential element in every site visited. Enforcement has played a large role in public acceptability by ensuring fairness because those paying for use of the road want to know that others are paying as well. Many pricing programs do not consider enforcement penalties as a revenue tool (i.e., fines and fees need not be higher than administrative costs), but do view enforcement as a critical element of ensuring that base road charges are collected without substantial leakage. All of the sites studied, except Stockholm, treat violations as administrative fees, not as criminal acts.

B. Violation enforcement systems require effective system integration and linkages with motor vehicle registries.

Typically, enforcement is managed through video capture of license plate images and an ANPR system. Back office processing centers use license plate information to identify the vehicle owner and collect payment.

In systems that rely on video systems and ANPR as the means of charging for road use, the enforcement process is an exception-based business process to pursue those who have not paid after some period of time. This process leverages one set of roadside equipment for dual functionality (i.e., primary collection and enforcement). This is the case in London and Stockholm.

In systems that rely on OBUs (i.e., transponders or GPS) for toll collection, the video enforcement process is a stand-alone subsystem that requires roadside video equipment for enforcement and systems integration to ensure violation transactions and electronic toll transactions are uniquely identified and properly distinguished. The systems in the Czech Republic, Germany, and Singapore rely on separate violations enforcement subsystems. All these systems supplement the automated violations enforcement systems with mobile and roadside enforcement efforts for periodic spot enforcement.

The use of ANPR enforcement requires an effective working relationship with motor vehicle registries for accurate information of vehicle owners from license plate data. Systems that include pricing for significant populations of foreign-registered vehicles (e.g., Germany and the Czech Republic) have a more complex set of relationships to establish and maintain in a cost- effective manner.

C. Performance standards ensure operational effectiveness through requirements in system design and procurement, as well as in service-level agreements for operations.

Performance standards cover a wide range of operating interests. For roadside operations, they often address system and roadway availability, traffic data capture requirements, and image reject rates of video equipment. For back office operations, typical performance standards include the error rate in invoicing, call center customer wait times, unprocessable image rates, and uncollectable transactions. The best practice in pricing programs is to establish service-level agreements with both external contractors and internal agency service providers to maintain operating performance with financial incentives.

Careful selection of performance standards is critical to ultimate outcomes. Systems that have emphasized accuracy and availability over financial performance have experienced high-cost operations with redundant systems and processes.

D. Best practice for back office operations depends on translating operating concepts into clear business rules and refining practices based on operating experience.

Business rules are the foundation of successful back office operations. Agencies involved in their development and refinement tend to gain advantages in managing operating performance and costs. For instance, Stockholm's move to a complete video-based system has resulted in 47 business rules governing the optical character recognition aspects of image processing, driving performance in accuracy and collections. Germany's decision to outsource the entire toll operation to Toll Collect has resulted in good system and operating performance, but relatively high operating costs.

Many programs have found the need to refine operating practices after implementation to manage costs more effectively. Over time, this has resulted in contractor changes in several pricing programs to best meet the operational and technical requirements at a competitive cost. London and Stockholm have both recently changed operators, while Germany and the Czech Republic face long-term contracts that provide less flexibility for managing operating costs more effectively.

Stockholm has also adopted a policy of insourcing that seeks to identify elements of the operation that may be managed with effective performance and lower costs by in-house resources rather than by outside services.

E. Payment channels and methods add operational and system complexity, but not always commensurate benefits.

During the congestion tax trial in Stockholm, vehicles without transponders had until noon the following day to pay the charge via the Internet or at a physical retail location, which included convenience stores, banks, and stand-alone kiosks. Charges not paid during the grace period were assessed a SEK70 (US$8) late fee for the first reminder and a SEK450 (US$62) fee for the second reminder. Since Swedish laws governing the collection of taxes required that the system capture all license plates as evidence of incurring the congestion tax, the transponder-based system was duplicative. The permanent Stockholm program phased out the transponder payment option and the associated costs of distribution and account maintenance. Adopting an all video-based system also created the opportunity to reduce the systems and accounting required for multiple postpayment channels. Instead, Stockholm placed the congestion pricing program and national vehicle registry in one organization that could take advantage of monthly invoices for the congestion tax. This reduced the cost of collection substantially and improved the collection rate.

Despite the many successes of the London congestion- pricing program, the high costs of operations have been often criticized. While transaction processing costs in a video-based system tend to be higher than those with OBUs (i.e., DSRC or GPS units), the multiple payment channels also add cost to the London system. Germany is also faced with high operating costs of its manual booking system, which requires more than one-third of its operating costs to handle 10 percent of the traffic volume. In Singapore, the payment method was simplified by standardizing in-vehicle units and designing a stored-value smart card interface that allows replenishment at banks and other outlets.

7. Outreach and Public Acceptance

A. Many forms of public involvement based on the cultural and political context of the host country were used to address public concerns about road pricing.

After several attempts to implement a distance charge, the Dutch realized that proactive stakeholder outreach during the planning and concept development stage is essential. Over the past 2 years, staff and leadership at the Dutch Ministry of Transport have invested heavily in public outreach and education. By engaging in a thorough and thoughtful planning and public involvement process, the Dutch developed clear, salient, and timely messages about the purpose and benefits of pricing, including "drive less, pay less."

Public outreach in Singapore included the opening of the LTA Transportation Gallery, an interactive exhibit that teaches younger audiences about concepts such as mobility, access, sustainability, land use, and demand management. The gallery is a powerful educational tool that explains the history, context, and future of ERP and other transportation options in Singapore.

B. Clear, salient, and timely messages about the purpose and benefits of pricing were used to help educate key stakeholders and garner public acceptance.

Singapore targets education about transportation solutions such as road pricing at youth with an interactive transportation gallery. Key messages used in Singapore include "keep roads free-flowing," "people-centered transportation," and "public transit is a viable choice."

Czech and German truckers supported pricing in an effort to "level the playing field" with foreign haulers and promoted the message of "user pays."

Based on prior false starts, the Dutch are investing heavily in stakeholder outreach and are committed to a revenue-neutral road-pricing scheme. Their mantra is "drive less, pay less."

C. Issues of equity and privacy were dealt with differently in each locale.

Equity issues related to a person's ability to pay the fee were not widespread in London, Singapore, or Stockholm because of the high cost of car ownership and existence of good transit alternatives in all three cities. Lower income commuters in these cities tend to use transit because the driving cost differential is significant. In addition to substantial transit investments, London and Stockholm provide a variety of exemptions and discounts to various users (i.e., transit vehicles, taxis, hybrids, monthly and annual pass purchasers, residents within the charging zone or in adjacent communities). Singapore, on the other hand, provides few exemptions (only for military and emergency vehicles). Because many transit providers in Singapore are privately contracted and operated, all transit vehicles are required to pay.

Privacy concerns were not as prevalent with the German and Czech truck toll systems because of the commercial nature of the enterprise. In Singapore, the privacy issue was dealt with through the use of an IU that requires a prepaid, stored-value smart card that did not hold any personal data. The smart card provides users with a high level of anonymity.

D. Short-term gains may negatively impact longer term program sustainability.

The western expansion of London's congestion charging zone was championed by Mayor Ken Livingstone and made effective in 2007. In 2010, under the administration of Mayor Boris Johnson, the western expansion was repealed after a series of public consultations.

Commitments from the German Parliament that HGV toll revenues would augment roadway funding were not kept, which jeopardizes future prospects to price other vehicles based on distance traveled or emissions class.

E. The focus on changing behavior through pricing is clear.

The Dutch Ministry of Transport's goal is to shift the cost from vehicle ownership to usage to create a more sustainable transportation system.

The Germans have adopted a user-pays principle for freight haulers. In addition, by having a graduated toll schedule for cleaner trucks, they have seen a 60 percent shift away from the Euro 1, 2, and 3 emission- level trucks to the cleaner Euro 4 and 5 emission-level trucks.

Singapore estimates that its gas tax would need to be raised by SG$3 to achieve the same traffic reduction that a SG$1 increase in its ERP system because of the transparency and direct price signal of the system.

5. Implementation Recommendations

Over 2 weeks in December 2009, the scan team engaged with practitioners, champions, and the foremost international experts on road pricing. In some respects, the host countries are akin to individual States or regions in the United States. Thus, from an implementation perspective the findings and lessons learned that form the basis for implementation ideas may be best suited for application at the State or regional level.

While all scan team members are integral to implementation, a six-person subset of the scan team, known as the Implementation Team, will facilitate efforts to advance the application of road pricing in the United States. The team, chaired by Patrick DeCorla-Souza, developed an Implementation Plan designed to capture and cultivate the key findings and lessons learned. The team recommends that additional resources and effort be focused on the following three strategic areas:

1. **Enhanced outreach and communications**—To advance the use of road pricing in the United States, it is paramount that transportation leaders, policymakers, key stakeholders, and a larger cross-section of the public understand the benefits and implications of broader road pricing.
2. **Additional research needs**—With the application of road pricing in the United States limited to HOT lanes, there is continued need for additional research to better

comprehend issues related to public perception, implementation barriers, behavioral effect, and integration of road pricing with multimodal land use and transit options.

3. **Road pricing toolkit**—Transportation professionals lack a comprehensive decision analysis tool to assess the merits of various road pricing options to address specific problems. The toolkit would include a module to assist in making design decisions, a guidebook or primer to assist technical managers in developing financing and procurement strategies, comprehensive and synergistic transportation plans that incorporate road pricing concepts applicable in the U.S. context, and analytical tools to estimate performance and costs of alternative concepts in comparison with conventional tax-based approaches. The tools would culminate in a decision tree tool to help transportation leaders make informed decisions on the relevance and feasibility of a road pricing alternative.

It will require strategic partnerships as well as social capital and funding to initiate the first phase of a technology and policy program that captures the best of the findings from the scan. The scan team will use the products developed in the efforts outlined to target selected States and regions that are the most likely candidates for the advanced road pricing approaches the scan team observed firsthand in the countries visited.

The following describes a series of foundational implementation items that would enable practitioners, elected officials, and other key stakeholders to advance the state of the practice in road pricing in the United States:

1. **Outreach to transportation agency leaders and policymakers**—This effort will involve targeted outreach to transportation leaders, policymakers, and various stakeholder organizations. This will include short briefs on how road pricing helps achieve various societal goals and a document providing responses to questions and concerns often made about road pricing.
2. **Messaging and public involvement**—This effort will focus on the psychology and messaging issues around pricing. It will include developing a more detailed understanding of the Dutch and Singaporean approach to communicating a radical change in paying for transportation to the public and stakeholders and translate it to the U.S. context.
3. **Behavioral implications and impacts of various pricing concepts**—This effort will include development of sketch-planning tools to estimate the travel behavior and consequent traffic, environmental, and economic impacts of various alternative pricing concepts in comparison with traditional methods of road funding. An example is the analytical approach used in Singapore to estimate that it would need a SG$3 gas tax increase to effect the same behavior change as a SG$1 increase in the ERP system.
4. **Road pricing decision tree**—This will provide a tool to assist transportation leaders in defining a road pricing scheme that best suits identified needs, socioeconomic climate, terrain and network constraints, and available technology and resources. It will address such issues as the specific purpose for pricing, potential negative impacts of change to the status quo, public acceptance, capital and operating costs, speed of implementation, potential for private participation, procurement alternatives, and technology options.

5. **Technology, procurement, and PPPs to implement pricing**—This effort will provide guidance on defining business and system requirements that are aligned with the stated objectives of a program that employs pricing as a fundamental element of its approach to transportation management and infrastructure improvement. It will build on an understanding of the procurement processes and financing techniques employed in the Czech Republic, Germany, and the Netherlands, including PPPs and other private sector roles in advancing pricing programs and the related infrastructure improvements and operations.
   This effort will attempt to develop guidance and specific applications for sustainable program successes in the U.S. context.
6. **Developing concepts applicable to the United States based on lessons learned**—Not all lessons learned on the scan are directly transferable to the U.S. context because of the differing political, social, urban development, and transportation mode characteristics in the areas visited. This effort will develop an array of alternative concepts based on the lessons learned that could potentially be applied in various U.S. contexts. These concepts would be promoted to targeted metropolitan areas and regions, potentially one or more of the areas represented on the scan team.
7. **Integrating pricing, land use, and transportation investment**—This effort will provide guidance on developing synergistic pricing, land use, and transportation policies and investment strategies.
   It will build on the approach used in Singapore to develop synergistic policies to create an efficient and effective transportation system by integrating demand management (including land use) and multimodal investment strategies.

APPENDIX A. SCAN TEAM MEMBERS

Robert (Bob) Arnold (FHWA Cochair)
Director, Office of Transportation Management
Federal Highway Administration
1200 New Jersey Ave. SE
Washington, DC 20590
Telephone: 202-366-1285
E-mail: robert.arnold@dot.gov

Vance C. Smith, Jr. (AASHTO Cochair)
Commissioner
Georgia Department of Transportation
600 W Peachtree St. NW
Atlanta, GA 30308
Telephone: 404-631-1005
E-mail: vsmith@dot.ga.gov

John Q. Doan (Report Facilitator)
Senior Associate
SRF Consulting Group, Inc.
1 Carlson Parkway, Suite 150
Minneapolis, MN 55447
Telephone: 763-475-0010
E-mail: jdoan@srfconsulting.com

Rodney N. Barry
Division Administrator, Georgia Division
Federal Highway Administration
61 Forsyth St., Suite 17T100
Atlanta, GA 30303
Telephone: 404-562-3630
E-mail: rodney.barry@dot.gov

Jayme L. Blakesley
Attorney-Advisor, Office of Chief Counsel
Federal Transit Administration
1200 New Jersey Ave. SE
Washington, DC 20590
Telephone: 202-366-0304
Fax: 202-366-3809
E-mail: jayme.blakesley@dot.gov

Patrick T. DeCorla-Souza
Program Manager, Office of Innovative Program Delivery
Federal Highway Administration
1200 New Jersey Ave. SE
Washington, DC 20590
Telephone: 202-366-4076
E-mail: patrick.decorla-souza@dot.gov

Mark F. Muriello
Assistant Director, Tunnels, Bridges, and Terminals Department
Port Authority of New York and New Jersey
One Madison Ave., 5th Floor
New York, NY 10010
Telephone: 212-435-4836
E-mail: mmuriello@panynj.gov

Gummada N. Murthy
Director, Operations and Security Division
Virginia Department of Transportation
1211 East Broad St., 4th Floor

Richmond, VA 23219
Telephone: 804-786-2978
E-mail: gummada.murthy@vdot.virginia.gov

Patty K. Rubstello
Toll Systems Development Engineer
Washington State Department of Transportation
WSDOT/Goldsmith Building
401 2nd Ave. S, Suite 300
Seattle, WA 98104
Telephone: 206-716-1299
E-mail: rubstep@wsdot.wa.gov

Nick A. Thompson
Director, Office of Policy Analysis, Research, and Innovation
Minnesota Department of Transportation
MS 670
395 John Ireland Blvd.
St. Paul, MN 55155-1899
Telephone: 651-366-3152
E-mail: nick.thompson@dot.state

Biographies

Robert E. Arnold (FHWA cochair) is the acting director of the Federal Highway Administration's (FHWA) Office of Transportation Management and director of the Office of Transportation Operations. These offices are responsible for national programs focused on reducing roadway congestion. The Office of Transportation Management deals with recurring congestion in such areas as pricing, active traffic management, and traveler information, while the Office of Transportation Operations has the nonrecurring side. This includes incident management, work zone operations, emergency transportation and organizational preparedness, the *Manual of Uniform Traffic Control Devices,* and various road weather management programs. As a member of FHWA's Senior Executive Service, Arnold is responsible for contributing to the development of overall strategic planning, policies, and legislative proposals for the administration. He was FHWA's New York Division administrator from February 2001 to September 2007. FHWA Divisions work primarily with State departments of transportation to ensure that each State's highway system is an integrated, effective, and efficient part of the national transportation system. Arnold has also held positions with FHWA as assistant division administrator in New Jersey, district engineer in New York, field operations engineer in Oklahoma, construction engineer in FHWA's Western Region, and area engineer in Baltimore, MD. He received a bachelor's degree in civil engineering from Ohio Northern University in 1983 and has worked for FHWA since graduation. Arnold is also a 2006 graduate of the U.S. Office of Personnel Management's Federal Executive Institute.

Vance C. Smith, Jr. (AASHTO cochair) is the commissioner of the Georgia Department of Transportation. He was elected to the Georgia House of Representatives in 1993 and has continuously served Georgia's 129th District. Smith has chaired the House Transportation Committee for the past 5 years and served as a member of the Appropriations, Economic Development, and Tourism and Rules committees. He was cochair of the General Assembly's 2007 Joint Study Committee on Transportation Funding. Smith is a graduate of Columbus State University with a bachelor's degree in business. He is a graduate of the Leadership Georgia program. Smith has held posts with the Council of State Governments, National Conference of State Legislatures, Southern Legislative Conference, and American Legislative Exchange Council.

John Q. Doan (report facilitator) is a senior associate at SRF Consulting Group in Minneapolis, MN. Doan leads SRF's road pricing practice with national and local expertise in transportation planning, innovative finance, public involvement, and intelligent transportation systems (ITS). At SRF, he helped develop the Minnesota Urban Partnership Agreement proposal and is the lead public outreach consultant for the Interstate 85 high-occupancy toll (HOT) lanes conversion project in Atlanta, GA. Before joining SRF, Doan worked in Wells Fargo's Public Finance Division, was assistant to the commissioner at the Minnesota Department of Finance, and most recently served as the MnPASS program director for the Minnesota Department of Transportation. He earned his bachelor's degree in civil engineering and public policy from Carnegie Mellon University and a master's degree in public policy from Harvard University's Kennedy School of Government. Doan is a registered engineer, member of the University of Minnesota Transportation and the Economy Research Council, SRF's liaison to the International Bridge, Tunnel, and Turnpike Association, and member of the Transportation Research Board's (TRB) Congestion Pricing Committee.

Rodney N. Barry is the division administrator of FHWA's Georgia Division. Barry directs a staff of 31 responsible for administering the Federal-aid highway program in Georgia. Since arriving in Georgia in 2007, he has worked with the Georgia Department of Transportation (GDOT) and various other State and local transportation agencies to address congestion in the Atlanta metropolitan area, one of the nation's most congested cities. He is assisting GDOT in implementing an HOT lane facility. Barry has been with FHWA for 23 years and held positions in North Carolina, Florida, Kansas, Nebraska, and FHWA headquarters in Washington, DC. He has bachelor's and master's degrees in civil engineering from Auburn University. He is a licensed professional engineer in North Carolina.

Jayme L. Blakesley is an attorney-advisor with the U.S. Federal Transit Administration (FTA) Office of Chief Counsel. He is FTA's lead attorney on road pricing, transit-oriented development, and public-private partnerships. Blakesley entered Federal service through the Presidential Management Fellowship Program, of which he is a graduate. Before joining FTA, he worked on transportation issues for the Office of the Utah Attorney General and the Utah Transit Authority. Blakesley has been recognized for work he performed to implement the National Strategy to Reduce Congestion on America's Transportation Network and for his significant contributions to the public transit industry. Blakesley is a graduate of Brigham Young University and has a law degree from the University of Utah's S.J. Quinney College of Law. He is a licensed member of the Utah State Bar. Blakesley is vice chair of the Alexandria (VA) Transportation Commission and a member of TRB's Committee on Transit and Intermodal Transportation Law.

Patrick T. DeCorla-Souza is the program manager for tolling and pricing in the FHWA Office of Innovative Program Delivery in Washington, DC. DeCorla-Souza has led U.S. Department of Transportation efforts to promote congestion pricing since 1999. He oversees congestion pricing initiatives across the United States, working with public and private sector partners to implement innovative road pricing strategies. DeCorla-Souza has developed new approaches to address public acceptance issues related to road pricing, new performance-based public-private partnership approaches for implementing pricing with complementary transit services, and simplified analytical tools that make it easier to provide critical impact information on road pricing strategies to decision makers. His ideas and research have been published in various U.S. and international professional and trade journals. He is the author of more than 100 technical papers and articles on road pricing, public-private partnerships, m ultimodal transportation evaluation using benefit-cost analysis, travel demand modeling, air quality analysis, and public involvement. DeCorla-Souza has master's degrees in transportation planning and civil engineering and cochairs the TRB Congestion Pricing Committee.

Mark F. Muriello is the assistant director of tunnels, bridges, and terminals for the Port Authority of New York and New Jersey. He is responsible for the six vehicular bridges and tunnels between New York City and New Jersey and two interstate bus terminals, which collectively serve 1.25 million customers each weekday. Muriello directs a wide range of functions, including transportation planning and policy, traffic and revenue programs, business and strategic planning, and operation and maintenance of the Port Authority's E-ZPass electronic toll collection system and time-of-day toll pricing program. He has 27 years of experience in transportation and public finance, covering toll, bus, rail, and marine terminal operations, as well as the electric utility industry. Muriello has a bachelor's degree in industrial engineering and operations research from Columbia University and a master of business administration in finance from New York University. Muriello cochairs TRB's Congestion Pricing Committee, represents the Port Authority on the E-ZPass Interagency Group's Policy Committee, chairs the Technical and Operations Committee for TRANSCOM (Transportation Operations Coordinating Committee), cochairs the Policy and Strategic Planning Committee of the I-95 Corridor Coalition, and is a member of the International Bridge, Tunnel, and Turnpike Association's Communications and Outreach Task Force.

Gummada N. Murthy is the director of the Operations and Security Division at the Virginia Department of Transportation (VDOT). He is responsible for developing and issuing statewide policy and procedures for all operations related to technology and ITS programs that support incident management, roadway weather information systems that support snow and ice removal programs, and VDOT's HOT and pricing programs. Murthy also has a responsible role in research and implementation of innovative technology in VDOT operations programs such as Global Positioning System (GPS) -based roadway information mining programs, and active traffic management and commercial vehicle operations. He served with VDOT for 3 years and has more than 25 years of experience in all phases of transportation planning and engineering with a focus on operations of roadway and roadside infrastructure, including ITS, signal systems, ITS deployment and operations, and roadway maintenance programs. Murthy has a bachelor's degree in civil engineering from the University of South Florida. He is a licensed professional engineer in Virginia and serves on several technical committees of the American Association of State Highway and Transportation Officials (AASHTO).

Patty K. Rubstello is the tolling and systems development engineer for the Washington State Department of Transportation (WSDOT). Rubstello is responsible for developing and implementing toll systems for the State of Washington. This includes conducting public education and outreach for, assessing the environmental effects of, and constructing toll systems. Over the past 7 years, she has managed several studies on how pricing could improve the operations of a number of highways in the Puget Sound region. In 2008, Rubstello implemented the first HOT in Washington State. She manages the implementation of pricing on an existing toll-free highway in the Seattle area. Rubstello's educational background is in mathematics and civil engineering. She is a licensed civil engineer in Washington. Rubstello participates in a number of pricing-related forums, such as the TRB Congestion Pricing Committee and the National Pricing Discussion Group hosted by WSDOT.

Nick A. Thompson is the director of policy analysis, research, and innovation for the Minnesota Department of Transportation (Mn/DOT), in St. Paul, MN. Thompson directs the implementation of Minnesota's Urban Partnership Program, which includes a project to expand Minnesota's congestion management HOT lane program, MnPASS, to its second transportation corridor. He has worked for Mn/ DOT for 13 years. Before his current position, he was the project manager for implementing Minnesota's first congestion pricing HOT lane, freeway operations manager, and area manager responsible for transportation project development in suburban and rural communities. He holds a bachelor's degree in geography from Gustavus Adolphus College and a master of urban planning degree from the University of Wisconsin-Milwaukee. Thompson has served as a member of TRB's Freeway Operations Committee, chaired the Transportation Management Center Pooled Fund Study, and served on the board of ITS Minnesota.

Appendix B. Scan Preparations and Itinerary

Scan Preparation

Planning for the road pricing scan trip began in February 2009 with a kickoff meeting that included original cochairs Peggy Catlin of the Colorado Department of Transportation and Regina McElroy of the Federal Highway Administration. The initial preparation culminated with the completion of a desk scan in April 2009 and an organizational meeting in Washington, DC, on April 30, 2009. The following were the objectives of the desk scan:

- Further the efforts of the full scan team in acquiring information of value to the U.S. transportation community.
- Increase the cost-effectiveness of the full scan by advising the team where best to commit its limited time abroad.
- Refine the scope of the scan by identifying relevant sources of information abroad and clarify the focus of the scan if it is determined to be too broad.

The scope of the desk scan was limited to office-based information gathering, designed to supplement and further refine the broader issues associated with implementing road pricing. The information that supported this desk scan originated from three primary sources: published literature, various Internet sites, and U.S. and international experts in the field.

A review of relevant published literature was conducted using the Transportation Research Information Services (TRIS) Online, with a particular emphasis on articles published in the last 10 years. In addition, other online tools such as Google, Wikipedia, and official road pricing program or project Web sites published by the sponsoring authority provided further project details. A list of relevant published literature and Web sites used for the desk scan is in Appendix C.

U.S. and international experts in the field of road pricing were identified based on the literature review, the report facilitator's contacts through the Transportation Research Board's (TRB) Congestion Pricing Committee and International Bridge, Tunnel, and Turnpike Association, and recommendations from other experts.

The desk scan recommended that the team visit Germany, the Netherlands, Singapore, Sweden, and the United Kingdom. After the desk scan was completed, a request was made to add meetings with representatives from the Czech Republic based on new information. While other countries, including Australia, Chile, Italy, Norway, and South Korea, were considered, the selected six countries could be visited within the constraints of a 2-week scan and provided the optimal combination of advanced practices and innovative implementation.

Team Meetings and Travel Itinerary

During the scan, the team visited representatives in five countries: Germany, the Netherlands, Singapore, Sweden, and the United Kingdom. Because of the logistical challenges of making a sixth country stop, representatives from the Czech Republic met with the scan team in London. The team members left the United States on Dec. 4, 2009, and held their first team meeting on Dec. 6 in Stockholm, Sweden. They met with Swedish representatives Dec. 7 and 8, and then flew to Berlin, Germany. Meetings with the German delegation occurred Dec. 9 and 10.

The first week ended in The Hague, Netherlands, with discussion on the upcoming Dutch implementation of a distance-based user fee on Dec. 11. The midpoint team meeting was held Dec. 13 in London, followed by meetings with British officials on Dec. 14 and the Czech delegation on Dec. 15. The team left the United Kingdom Dec. 15 and arrived in Singapore on Dec. 16. Meetings with representatives in Singapore occurred on Dec. 17 and 18, and the wrap-up team meeting was held Dec. 19 in Singapore.

The team developed a scan summary presentation and shared it with various TRB committees and subcommittees at the annual TRB Conference in Washington, DC, in January

2010. A summary report as well as drafts of the final report and implementation plan were written and produced for distribution at the team's reassembly meeting in Washington, DC, on March 10 and 11. The following chart summarizes the team meetings and travel schedule.

Date	Location	Date Location Purpose or Hosts
April 30, 2009	Washington, DC	Initial team meeting to determine emphasis areas and countries to visit and develop amplifying questions
Dec. 6, 2009	Stockholm, Sweden	Kickoff trip meeting to review travel plans, set personal scan goals, and assign roles
Dec. 7–8, 2009	Stockholm, Sweden	Meeting with Swedish Road Administration, Traffik Stockholm, and City of Stockholm
Dec. 9–10, 2009	Berlin, Germany	Meeting with Federal Ministry of Transport, Building and Urban Development, Federal Infrastructure Finance Group, Toll Collect GmbH, and German Trucking Association
Dec. 11, 2009	The Hague, Netherlands	Meeting with Ministry of Transport, Public Works, and Water Management
Dec. 13, 2009	London, United Kingdom	Midtrip meeting to review findings to date
Dec. 14, 2009	London, United Kingdom	Meeting with British Department for Transport and Transport for London
Dec. 15, 2009	London, United Kingdom	Meeting with Czech delegation from the Ministry of Transport and Czech Technical University
Dec. 17–18, 2009	Singapore	Meeting with Ministry of Transport and Land Transport Authority
Dec. 19, 2009	Singapore	Final trip meeting to identify key findings and develop preliminary recommendations

Appendix C. Amplifying Questions

General/Background

1. What is the financial health of transportation in the country? Are there sufficient funds to meet needs? Have there been any major changes in funding/costs in the last 10 years, and what does the future look like for revenue?
2. Beyond tolling, pricing, and fare revenue sources, how are transportation in general and highways specifically funded?
3. Other than monetary fees, have other incentives/ disincentives been used to change user behavior?
4. Did general/traditional tolling exist before the variable pricing concept was developed?
5. Did you experience any setbacks or outright failures during the project development process? If so, what were the causes and what would you do differently?

Concept Development and Planning

1. Where did the idea for this project come from? How long did the idea take to go from concept to operation?

2. What was the more important driver in launching this program—politics, transportation funding shortfalls, congestion levels, or environmental issues? Has the focus changed over time on why this is needed?
3. How are "pricing" and "tolling" concepts defined and distinguished—at the policy/program level and at the network application/operation level?
4. How was the pricing program developed? Did the cost estimates for such projects include operation and maintenance (O&M) costs in addition to the capital costs? Was pricing structure validated to generate theoretical revenues that guaranteed to support such functions?
5. What is the relationship between road pricing as a revenue stream and an operational strategy to reduce congestion, improve safety, or enhance the environment?
6. What are the relevance and importance of including supporting strategies such as transit enhancements, system operations strategies and technologies, and travel demand management?
7. Are there plans to expand the system or add different types of congestion pricing?
8. What is the concept of operations for tolling and pricing—at the project scoping level through operations— encompassing policy, strategic, technology, and interagency elements with what-if scenarios built in (if any)?

Political, Organizational, and Institutional Issues

1. What was the purpose of the pricing initiative?
2. Who is responsible for the overall program? How are the transportation responsibilities segmented (local, national, private)?
3. What are private and public relationships in the design, construction, and operations of the system? For public-private partnerships (i.e., concessions, if any) what resources were provided by the agency vis-à-vis the private partner?
4. What were the political challenges that had to be overcome to create the program?
5. Were transit improvements required as part of a package to get acceptance for the idea? If so, what were the components?
6. Were any legal changes needed to implement the system?
7. What political barriers had to be addressed to gain acceptance of the pricing initiative? What strategies were used to address political history and regulatory and legal barriers, particularly those related to public acceptance?
8. What was the level of mainstreaming pricing/tolling into the institutional planning, construction, and operations programs?
9. What institutional arrangements and interagency collaborations were needed to enable effective applications of pricing techniques? Were there institutional barriers or interagency conflicts that had to be overcome?

Pricing Policy

1. Is maximizing revenue, minimizing congestion, or a combination a goal of the pricing scheme?

2. What types of users/vehicles are exempt from paying the toll or receive a discount? How do the prices differ for local residents and nonresidents?
3. Was the pricing initiative on a new or existing transportation facility?
4. How is pricing set? What factors drive the price—day, time of day, level of congestion, distance, emissions standards, or environmental conditions?

Costs and Revenues

1. How much did it cost to implement your pricing program (capital and ongoing)?
2. How much revenue does the system generate? Does it pay for capital and operating costs?
3. Do revenues meet expectations projected before the program launched? Are revenues adequate to support targeted service levels? What's the safety cushion for unforeseen shortfalls?
4. How are net revenues (after debt and O&M cost are covered) distributed (to other transportation projects, other modes, nontransportation uses, etc.)? Is it legislatively mandated?
5. Has the program achieved economies of scale yet?
6. Does the private sector share in the revenue? If not, was this considered?
7. How are revenues incorporated into the agency budgeting and program development process?
8. How are revenues faring in the current economic climate?

Tolling Technology and Roadway Design

1. What technology is used for revenue collection and enforcement? What are the technical details of the pricing system?
2. How did the revenue collection system evolve (standardized and then implemented, independent systems, or a mixture of systems that have been cobbled together into a single back office)? Have there been any major evolutions or changes in the pricing system since the start of the program?
3. Did implementing pricing require any specific design changes for geometrics, signing, etc., or modification of the legal code? If applicable, at the termini of the pricing project/area, how are the upstream and downstream connections made (i.e., going from a priced to nonpriced facility)?

Operations

1. What technologies for operations were used, such as traditional intelligent transportation systems solutions and their expansive application to pricing and tolling applications, lane control dynamic message signs for variable pricing, active traffic management, or integrated weather and actionable roadway condition reporting systems to manage weather and other nonrecurring events?

2. How do the back-office tolling operations interface with other roadway management operations (incident clearance, freeway monitoring, dispatch, etc.)?
3. Are there cases when the tolls are waived or turned off, such as under extreme events or systemwide requirement? Who makes that decision?
4. Is there any delineation of the operations and maintenance service level for priced and nonpriced facilities?
5. How is interoperability of toll systems handled?
6. How does your enforcement system work? Who does the enforcement, what are the fines, and is it meeting your goals? What is your violation rate and who are the typical cheaters?
7. What has been the operational experience of pricing and/or tolling on freight haulers in terms of route diversion and time-of-day use? Other than the actual amount charged, are commercial freight haulers handled differently under the pricing schemes?
8. What transit alternatives are available? Were additional transit alternatives provided as a part of the project?
9. How long did it take for the pricing program to reach an operational phase that began to deliver tangible benefits?

Evaluation of Impacts and Performance Measurement

1. What performance measures do you have? How do you report them?
2. Have you been able to identify any economic impact (positive and negative) on consumption-oriented businesses (sales, delivery charges, inventory control, staffing, etc.) caused by pricing?
3. What has been your experience with quantifying projected and actual benefits (e.g., reduction of congestion, safety, and environmental) and developing performance metrics for them?
4. What has been the effect of pricing on adjacent nonpriced facilities or areas? What have been the boundary effects?

Public Acceptance and User Issues

1. How have users and nonusers reacted to your pricing program and its associated technology? What was the initial public reaction? Has it changed over time?
2. What successful or unsuccessful techniques/strategies were developed and employed to gain acceptance of pricing and tolling programs?
3. Did the public have any privacy concerns and, if so, how were they mitigated?
4. What, if any, were the socioeconomic concerns of pricing and what way, if any, did you address them?
5. Are users required to have transponders or preregister to use the facility?
6. Who were your key champions and project spokespeople?
7. What were your key messages and how did you develop them?
8. Did you do any market research or consumer studies before or after implementation?

APPENDIX D. LITERATURE REVIEW AND INTERNET SOURCES

Published Literature

[1] Bhatt, Kiran, Higgins, Thomas, & Berg, John T. (2008). *Lessons Learned from International Experience in Congestion Pricing.* K.T. Analytics, Inc., Federal Highway Administration.

[2] (2007). CURACAO. *Work Package II: State of the Art Report (Draft).* Coordination of Urban Road User Charging and Organizational Issues, University of Leeds for the EC Curacao Project, United Kingdom.

[3] DeCorla-Souza, Patrick. (2008). *The Pricing's Right: Examining High-Performance Highways.* Auto Intermediates Limited.

[4] (2007). Dutch Ministry of Transport, Public Works, and Water Management. *Making a Start on a Price per Kilometre.* December.

[5] Eggers, William, et al. (2003). *Road User Pricing Could Help Ease and Manage International Traffic Congestion.* Deloitte Research, Novembe.

[6] Federal Highway Administration. (2006). *Congestion Pricing—A Primer.* December.

[7] Floyd, Doug. (2008). *Road Pricing in an Urban Context.* Transportation Association of Canada Briefing, Urban Transportation Council, February.

[8] Hin, Leo Tan Wee, & Subramaniam, R. (2006). *Congestion Control of Heavy Vehicles Using Electronic Road Pricing: The Singapore Experience.* International Association for Vehicle Design/ Inder Sciences Enterprises Limited.

[9] (2007). *ITS International Journal.* "Island Solution—Birger Hook, Project Director of Stockholm's Congestion Charging Trial, Talks about the Moves to Make the Scheme Permanent." ITS NA Edition, July/August.

[10] May, Anthony D., & Sumalee, A. (2005). *One Step Forward, Two Steps Back? An Overview of Road Pricing Applications and Research Outside the United States.* Transportation Research Board.

[11] Replogle, Michael. (2008). *Is Congestion Pricing Ready for Prime Time?* American Planning Association.

[12] Robinson, Ferrol. (2006). *Pricing Experience in Northern Europe: Lessons Learned and Applicability to Minnesota and the United States.* Humphrey Institute of Public Affairs, University of Minnesota, October.

[13] Robinson, Ferrol. (2008). *Heavy Vehicle Tolling in Germany: Performance, Outcomes and Lessons Learned for Future Pricing Efforts in Minnesota and the United States.* Humphrey Institute of Public Affairs, University of Minnesota, October.

[14] (2005). *Stockholm Moves To Beat Congestion.* Route One Publishing Limited.

[15] Thorpe, N. (2002). *Public Acceptance Of Road-User Charging: A Case-Study Of The Toll Rings In Norway.* International Association of Traffic and Safety Sciences.

[16] Wachs, Martin. (2005). *Then and Now: The Evolution of Congestion Pricing in Transportation and Where We Stand Today.* Transportation Research Board.

[17] Willoughby, C. (2000). *Singapore's Experience In Managing Motorization and Its Relevance To Other Countries.* World Bank.

[18] Woodroofe, Ian. (2006). *Continental Shift: Road Tolling Across Europe is Changing.* Auto Intermediates Limited.

Web Sites

Singapore
Land Transport Authority, www.lta.gov.sg/
http://en.wikipedia.org/wiki/Electronic_Road_Pricing
http://en.wikipedia.org/wiki/Singapore
Scheme

United Kingdom
Transport for London, www.tfl.gov.uk/
http://en.wikipedia.org/wiki/London_congestion_charge

Sweden
Swedish Transport Agency, www.transportstyrelsen.se/en/
http://en.Wiki pedia.org/wiki/Stockholm_congestion_tax

Germany
Federal Ministry of Transport,
Building, and Urban Development,
www.bmvbs.de/en
http://en.wikipedia.org/wiki/Toll_Collect
http://emagazine.credit-suisse.com/app/article/index.cfm
http://www.toll-collect.de/frontend/Homepage

Czech Republic
Ministry of Transport, www.mdcr.cz/en/HomePage.htm

APPENDIX E. HOST COUNTRY CONTACTS

Sweden

Swedish Road Administration
Fredrik Friberg
International Strategist
Swedish Road Administration
Vägverket
Röda vägen 1
SE-781 87 Borlänge, Sweden
Telephone: 011+46 243 752 56
Mobile: 011+46 70-398 51 27
E-mail: fredrik.friberg@vv.se

Germany

Federal Ministry of Transport, Construction, and Urban Development
Ulfert Joop
Robert-Schuman-Platz 1
D53175 Bonn (Germany)
Telephone: 011+49 228 300 5141
E-mail: joop@bmvbs.bund.de

Transport Infrastructure Finance Corporation
Stefan Buitkamp
Georgenstrabe 25
10117 Berlin, Germany
Telephone: 011+49 30 520 02 6221
Fax: 011+49+30+520 02 6212
E-mail: stefan.buitkamp@vifg.de

Toll Collect
Martin Rickmann
Toll Collect GmbH
Linkstraße 4
10785 Berlin, Germany
Telephone: 011+49 (0)30 74077-2400
E-mail: martin.rickmann@toll-collect.de

The Netherlands

Ministry of Transport, Public Works, and Water Management
Monique van Wortel
Senior Policy Advisor, ABvM/Road Pricing Project
Plesmanweg 1-6 PO Box 20901
2500 EX The Hague, Netherlands
Telephone: 011 +06 525 96 940
E-mail: monique.van.wortel@minvenw.nl

United Kingdom

Department for Transport
Neil Schofield
Department for Transport Zone 3/01
Great Minster House
76 Marsham St. London SW1P 4DR
Telephone: 011+44 300 330 3000
E-mail: neil.schofield@dft.gsi.gov.uk

Transport for London
Maxwell Forwood
Visits Executive
13th Floor, Windsor House
50 Victoria St.
London SW1H 0TL
Telephone: 011+020 7126 4557
Fax: 011+020 7126 4972
E-mail: maxwellforwood@tfl.gov.uk
Web site: www.tfl.gov.uk

Czech Republic

Ministry of Transport
Vaclav Cerny
Head of Unit, Telematics and Road Network Charging
Nábřeží Ludvíka Svobody 1222/12
110 00 Praha 1, Czech Republic
Telephone: 011 +420 225 131 241
E-mail: vaclav.cerny@mdcr.cz

Czech Technical University in Prague
Ladislav Bina (Assoc. Prof., Ing., CSc.)
Faculty of Transportation Sciences
Department of Logistic and Transportation Processes
Horska 3, 128 03 Praha 2, Czech Republic
Postal Address: Konviktska 20, 110 00 Praha 1
Telephone: 011+420 234 359 175
Fax: 011+420 224 919 017
E-mail: bina@fd.cvut.cz
Web site: www.fd.cvut.cz

Singapore

Land Transport Authority
Lim Lay Kim
Corporate Marketing
No. 1 Hampshire Rd.
Singapore 219428
Telephone: 011+ 65 6396 1597
Fax: 011+65 63961650
E-mail: lay_kim_lim@lta.gov.sg

Land Transport Authority
Hisam Hasim
Corporate Marketing
No. 1 Hampshire Rd.
Singapore 219428
Telephone: 011+65 6396 1641
Fax: 011+65 6396 1650
E-mail: hisam_hasim@lta.gov.sg

APPENDIX F. CURRENCY EXCHANGE RATES

The following table illustrates the exchange rates between the official currency of the host country for each scan site and the U.S. dollar. These exchange rates are certified by the Federal Reserve Bank of New York based on data collected from a sample of market participants. The exchange rates are for Dec. 12, 2009.

Currency	Exchange rate	
	USD per Unit	Unit Per USD
United Kingdom Pound (GBP)	1.6234	0.6159
European Union Euro (EUR)	1.4514	0.6891
Czech Republic Koruna (CZK)	0.0566	17.6599
Swedish Krona (SEK)	0.1392	7.1829
Singapore Dollar (SGD)	0.7172	1.3943

Source: Federal Reserve Statistical Release, March 17, 2010. Foreign Exchange Rates, Historic Data, www.federalreserve.gov/RELEASES/H10/Hist/.

INDEX

D

E

F

G

H

I

J

L

M

N

O

P

Q

R

S

T